GIS Tutorial

for Homeland Security

SUSAN LINDELL RADKE

EDDIE HANEBUTH

ESRI PRESS
REDLANDS, CALIFORNIA

ESRI Press, 380 New York Street, Redlands, California 92373-8100

Copyright © 2008 ESRI
All rights reserved. First edition 2008

10 09 08   1 2 3 4 5 6 7 8 9 10

Printed in the United States of America

Library of Congress Cataloging-in-Publication Data
Radke, Susan Lindell, 1957–
     GIS tutorial for homeland security / Susan Lindell Radke, Eddie Hanebuth.—1st ed.
       p.   cm.
     Includes bibliographical references.
     ISBN 978-1-58948-188-6 (pbk. : alk. paper)
     1. Geographic information systems. 2. National security.—United States. I. Hanebuth, Eddie, 1960– II. Title.

G70.212.R33 2008

363.340285—dc22   2008026439

Ask for ESRI Press titles at your local bookstore or order by calling 1-800-447-9778. You can also shop online at www.esri.com/esripress. Outside the United States, contact your local ESRI distributor.

ESRI Press titles are distributed to the trade by the following:
In North America:
Ingram Publisher Services
Toll-free telephone: 1-800-648-3104
Toll-free fax: 1-800-838-1149
E-mail: customerservice@ingrampublisherservices.com

In the United Kingdom, Europe, and the Middle East:
Transatlantic Publishers Group Ltd.
Telephone: 44 20 7373 2515
Fax: 44 20 7244 1018
E-mail: richard@tpgltd.co.uk

Cover design	Elizabeth Davies
Interior design	Donna Celso
Copyediting	Julia Nelson
Print production	Cliff Crabbe & Lilia Arias

Contents

Preface

In the wake of September 11, 2001, and Hurricane Katrina, much attention has been focused on how to provide our communities and agencies with the most effective tools and solutions to prepare for, protect against, respond to, and recover from threats to our national security. To that end, *GIS Tutorial for Homeland Security* provides readers a broad-based, hands-on resource for learning how to use geographic information system (GIS) technology in homeland security planning and operations. It includes a series of exercises that integrate the best practices of GIS and public safety to safeguard the nation in times of deliberate attacks and natural disasters.

This tutorial complements ESRI Press' recent publication, *GIS for Homeland Security*, written by Mike Kataoka. With a minimum of technical language, *GIS for Homeland Security* chronicles the accomplishments of progressive programs that have embraced geospatial technology to assess vulnerabilities, coordinate response, and facilitate recovery. *GIS Tutorial for Homeland Security* reaches a step further, providing hands-on instruction on how to perform the essential GIS tasks incorporated into homeland security planning and operations.

Each chapter in this tutorial integrates GIS analysis with a National Planning Scenario developed by the U.S. Department of Homeland Security (DHS). The scenarios selected in this tutorial include biological and chemical attacks, and natural disasters. The exercises apply geospatial analysis to enhance situational awareness, disseminate vital information, protect critical infrastructure and populations, and help communities recover from destructive and devastating events. It is by no means intended to address all potential threats to our national security, be they deliberate attacks or natural disasters. What is contained is a sampling of possible scenarios within our borders in the near or distant future. Regardless of the exact nature of the threat, the GIS tools and analysis presented here can be applied across the range of scenarios in whatever guise they occur.

The target audience for this tutorial is emergency service professionals new to GIS. It can be an effective teaching tool, both in a classroom setting focused on specialized training or as a self-study guide for individual professional development. An accompanying DVD contains PowerPoint presentation slides summarizing the essential components of each exercise. An instructor resource document outlining possible lesson plans and course schedules is in the book and on the DVD. These resources are designed to enhance any instructional setting, be it a classroom or individual workstation.

If you are new to GIS, it is important to work through each chapter in order. If you are an experienced GIS user, chapter 2 would be review of basic ArcMap components. Chapter 3 is the primary building block for later exercises, so it is essential. While the remaining chapters focus on independent emergency scenarios, they do cover the wide range of skills that encompass a comprehensive use of GIS analysis and tools. They can be selectively completed, though the process of homeland security planning and operations includes a definitive linear progression from preparation to prevention, to protection to response, and finally, to recovery. However unlikely that a particular event, such as a hurricane, would occur in your region, it is important to work through the exercises. You will learn to apply GIS tools and analysis to emergencies more likely to occur in your region. In the end, it is suggested that you complete the exercises in sequence to realize the full breadth of the homeland security planning and operations process.

The DVD included with this tutorial contains data for the chapter exercises. A separate DVD contains a 180-day trial version of ArcGIS 9.3. Instructions for installing the discs are provided in appendix B.

It is with sincere hope, and a sense of faith, that I embrace and deliver this technology as a most helpful tool to combat the threats that challenge the safety, security, and liberties of our nation.

Susan Lindell Radke
Orinda, California

Acknowledgments

We thank the professionals and organizations that supported the exercises in this tutorial by supplying data and advice regarding the use of GIS tools and analysis in their areas of expertise. The U.S. Geological Survey (USGS) has done extensive work preparing and presenting geospatial data and analyses to support homeland security planning and operations. Much of the data used in this tutorial is in the public domain, made accessible by the USGS via its data portals, viewers, and special programs Web sites. We also thank the U.S. Department of Homeland Security for integrating geospatial technologies into national strategies.

We appreciate the many organizations and individuals whose support and expertise are mirrored in the exercises. Special thanks to Tara Burkey of the U.S. Army Space and Missile Defense Command/U.S. Army Forces Strategic Command (SMDC/ARSTRAT) Measurement and Signature Intelligence/Advanced Geospatial Intelligence (MASINT/AGI) Node; Brian Quinn of the City of Berkeley; Jon Heintz of the Applied Technology Council (ATC); Diane Vaughan of the California Geological Survey; GIS specialist Gordon Ye; programming specialist Vladimir Ulyashin; ESRI utility specialist Bill Meehan; and John Radke, associate professor of environmental planning and geographic information science at the University of California, Berkeley. Thanks also to Dré Helms, Liz Rotzler, and Austin Smith of Digital Quest Inc.

The ESRI Press staff, including Peter Adams, Judy Hawkins, Mike Kataoka, Donna Celso, David Boyles, Savitri Brant, Julia Nelson, Jennifer Hasselbeck, Kathleen Morgan, Kelley Heider, and Jay Loteria provided tremendous support and expertise in bringing this tutorial book to fruition. Special thanks to ESRI Press cartographer Michael Law, who helped fine-tune the exercises.

We are especially grateful to ESRI President Jack Dangermond, who has always been receptive to our ideas.

I dedicate this book to John, who taught me to ask more intelligent questions.—*S.L.R.*

I dedicate this book to my staff, my friends, and my family.—*E.H.*

About the authors

 Susan Lindell Radke is founder and president of Berkeley Geo Research Group, an Orinda, California, GIS company. She has more than 25 years of practical experience in the design and development of GIS projects and geospatial information management systems, curriculum materials, database programming, computer network systems, environmental planning, and geography education. Ms. Radke has developed course material for schools, colleges, and professional development, encompassing primary geographic analysis, health and public safety, agricultural resource survey and land suitability analysis, homeland security, law enforcement, and economic development.

 Eddie Hanebuth is founder and president of Digital Quest, a Mississippi-based development and training-oriented company that produces GIS instructional material for educational institutions. He chairs the U.S. Department of Labor's National Standard Geospatial Apprenticeship Program and the SkillsUSA Geospatial Competition Committee, and runs the SPACESTARS teacher-training laboratory in the Center of Geospatial Excellence, NASA's John C. Stennis Space Center.

Notes for instructors

This book is a series of hands-on learning exercises covering GIS applications in the discipline of homeland security. It is a guide that supports the U.S. Department of Homeland Security (DHS) initiatives driving the government's efforts to safeguard the country. This tutorial parallels the target initiative of the National Preparedness Goal by providing a sampling of GIS applications that directly support mission area efforts to prevent, protect against, respond to, and recover from major events that affect our national security.

Chapter 1 introduces the concept of GIS and how it can be applied to homeland security operations. It outlines the primary initiatives in support of the DHS, such as the National Preparedness Goal, National Planning Scenarios, Target Capabilities List, and the Universal Task List. The chapter also provides an overview of the ArcGIS user interface.

Chapter 2 introduces ArcMap's suite of navigation and viewing tools and explores the layers relevant to homeland security operations. Students will learn to manage geospatial data in map and tabular formats as well as use tools helpful in navigating a map display.

Chapter 3 tells how to prepare a comprehensive dataset with geospatial data layers that are critical to the safety and security of the nation. Students will explore and build the components of a Minimal Essential Data Set, as defined by DHS. Portions of this data will be used in later exercises to address the prevention, protection, response, and recovery of communities in crisis scenarios.

Chapter 4 extends the basic ArcMap skills learned in earlier exercises to the preparation and production of reports and map layouts. Within the *Prevent* Mission Area of homeland security planning, students will learn to prepare warning reports and layouts essential to disseminating critical information that could prevent an attack.

Chapter 5 applies the suite of ArcMap tools and geospatial analysis to the *Protect* Mission Area of homeland security planning and operations. In this chapter, students will focus on two national planning scenarios to demonstrate how GIS tools and analysis aids and enhances the protection of the nation's people and places.

Chapter 6, in a *Respond* Mission Area scenario, brings students to the aftermath of Hurricane Katrina along the Gulf Coast, where emergency response converged to locate and rescue missing persons. Students will develop skills such as geocoding and preparing suitability maps in this scenario.

Chapter 7, the final chapter of this tutorial, applies GIS tools and analysis to the *Recover* Mission Area of homeland security planning and operations. Students will use USGS ShakeMaps to predict where earthquake devastation is most likely to occur and coordinate efforts with appropriate parties to restore normalcy in a particular area. Recovery includes immediate action to restore essential services to a community after a major disaster or event.

The data DVD includes PowerPoint presentations for chapters 1 through 7, summarizing the exercise content and, if applicable, providing an introduction to the chapter case scenario.

Note that the trial software packaged with the book expires after 180 days. For this reason your students may find it necessary to perform the exercises in a university lab with ArcGIS software.

If you are an instructor at an institution with a campuswide ESRI site license, you may order ArcGIS One-Year Time-Out Student Edition software for your students. Go to www.esri .com/slpromo for more information.

Notes for instructors
Chapter and related course scenario GIS application

Chapter	Scenario	GIS Application
1. Introduction to GIS and homeland security planning and operations	Students will be introduced to terminology and concepts relevant to GIS technology and will gain experience with the ArcMap interface. In addition, this chapter sets the framework for future chapters by introducing initiatives of the U.S. Department of Homeland Security (DHS) and outlining how GIS technology supports and enhances these initiatives.	• Learning what GIS is • Learning how GIS supports homeland security operations • Introducing the ArcGIS user interface
2. Visualizing data for homeland security planning and operations	Students will explore ArcMap navigational tools and techniques within the context of data layers that are relevant to homeland security operations.	• Manipulating map layers in a map • Viewing map layer properties • Viewing map layer attribute tables • Using zoom and pan tools • Creating spatial bookmarks • Using magnifier window • Using overview window • Using viewer window • Measuring distance • Identifying features • Finding features
3. Compiling data for homeland security planning and operations: *Prepare*	Students will explore geographic data that makes up the Minimum Essential Data Set (MEDS) as defined by DHS. Activities involve the use of various online sources to acquire, process, and manipulate data of differing formats into usable, relevant data as outlined by DHS.	• Using a new empty map document • Delineating DHS urban area • Adding boundary data to the MEDS map • Adding transportation data to the MEDS map • Adding hydrography data to the MEDS map • Adding structures data to the MEDS map • Adding a raster image layer to the MEDS map • Adding raster land cover layer to the MEDS map • Adding a DEM layer to the MEDS map • Adding geographic names to the MEDS map • Managing MEDS data

Notes for instructors continued on next page.

Chapter	Scenario	GIS Application
4. Designing map layouts for homeland security planning and operations: *Prevent**	Students will explore GIS tools and processes within the context of DHS National Planning Scenario 5: Chemical attack—blister agent.	• Preparing scenario map • Preparing warning report • Building warning map layout • Learning GIS outputs
5. Analyzing data for homeland security planning and operations: *Protect**	Students will explore GIS tools and processes within the context of: 1. DHS National Planning Scenario 12: Explosive attack—bombing using improvised explosive devices 2. DHS National Planning Scenario 6: Chemical attack—toxic industrial chemicals	• Preparing prevent scenario map 1 • Locating critical infrastructure • Protecting critical infrastructure • Preparing prevent scenario map 2 • Locating impacted population • Protecting impacted population
6. Analyzing data for homeland security planning and operations: *Respond**	Students will explore GIS tools and processes within the context of DHS National Planning Scenario 10: Natural disaster—major hurricane.	• Preparing response scenario map 1 • Geocoding missing person database • Preparing response scenario map 2 • Performing response suitability analysis • Creating suitability model using ModelBuilder • Adding the U.S. National Grid
7. Analyzing data for homeland security planning and operations: *Recover**	Students will explore GIS tools and processes within the context of recovery activities including: 1. Restoration of lifelines 2. Damage assessment	• Preparing recover scenario map 1 • Preparing recover scenario map 2

* Chapters 4-7 contain the Department of Homeland Security scenarios that focus on the use of GIS in the homeland security mission areas of prevention, protection, response, and recovery.

Instructors can use this tutorial in one of two scenarios to teach GIS applications in homeland security.

Scenario 1: Application resource for a course in GIS applications in homeland security

In this scenario, a college or university would provide a course offering in GIS applications for homeland security. *GIS Tutorial for Homeland Security* would supply hands-on, computer-based applications for the course. A supplemental textbook in homeland security could be adopted to include additional context into how GIS technology can be a valuable tool in homeland security administrative decision making and front-line problem solving. The choice of the supplemental text should be based on the focus that the department wants to adopt for its course offerings. A sample textbook will be used for the sample course schedules that are provided in this documentation. Sample course schedules are included for course arrangements with the following course length and weekly contact hour structures: 16 weeks/3 hours per week, 8 weeks/6 hours per week, 10 weeks/4.5 hours per week, and 5 weeks/9 hours per week.

Scenario 2: Professional development tool for homeland security administrators, first responders, and GIS professionals

In this scenario, professionals in the fields of homeland security, emergency management, protective services, municipal administration, geospatial technology, and others could use this text as a professional development tool in a formal training environment or as a self-study resource. Professionals in the fields of homeland security, emergency management, protective services, and municipal administration would be more familiar with the scenarios that are described in the text and would use this text to learn more about geospatial tools that could enhance their job performance. On the other hand, GIS professionals who may already be very familiar with the tools and techniques described in this book would have the opportunity to broaden their understanding of how these tools and techniques are applied in practical circumstances relating to homeland security.

The following information offers additional detail for using *GIS Tutorial for Homeland Security* in each of these scenarios.

Scenario 1: Application resource for a course in GIS applications in homeland security course

In this scenario, *GIS Tutorial for Homeland Security* serves as the primary resource for hands-on activities involving the application of GIS in the field of homeland security. It is strongly recommended that a secondary text be used as a resource for providing more extensive information about common considerations relevant to homeland security practices. The combination of texts would provide a more comprehensive approach to the course and theoretical and practical information useful to course participants.

A sample text is provided as a possible supplemental resource for the course.

Kataoka, Mike 2007. *GIS for Homeland Security*. ESRI Press, Redlands, CA. ISBN: 978-1-58948-155-8

Other resources can also be integrated into the course to further enhance the text materials. The following materials are readily available and can be found by using any common search engine.

Document	Description
National Preparedness Guidelines	This document created by DHS outlines the National Preparedness Goal, the National Planning Scenarios, the Universal Task List, and the Target Capabilities List in accomplishing national preparedness for hazards.
National Infrastructure Protection Plan Sector-Specific Plans	Prevention- and protection-focused reports from DHS that outline critical infrastructure and key resources.
National Response Framework	This document provides a guide to how the country responds to all-hazards incidents.

The following table provides a sample schedule for the planning of instructional resources for a course designed with a 16-week/3 contact hours per week structure.

<table>
<tr><td colspan="3" align="center">**GIS Tutorial for Homeland Security**
Proposed timeline
16 weeks, 3 contact hours per week</td></tr>
<tr><td colspan="3">*Resources*
Kataoka, *GIS for Homeland Security*
Radke and Hanebuth, *GIS Tutorial for Homeland Security*</td></tr>
<tr><td>**Date**</td><td>**Text chapter**</td><td>***GIS Tutorial for Homeland Security* chapter**</td></tr>
<tr><td>Week 1</td><td>*GIS for Homeland Security*, chapter 1</td><td>Chapter 1   Introduction to GIS and homeland security</td></tr>
<tr><td>Week 2</td><td>*GIS for Homeland Security*, chapter 1</td><td>Chapter 2 Visualizing data for homeland security options</td></tr>
<tr><td>Week 3</td><td></td><td></td></tr>
<tr><td>Week 4</td><td>*GIS for Homeland Security*, chapter 5</td><td>Chapter 3   Compiling data for homeland security planning and operations: Prepare</td></tr>
<tr><td>Week 5</td><td></td><td></td></tr>
<tr><td>Week 6</td><td></td><td></td></tr>
<tr><td>Week 7</td><td></td><td>Conclude chapter 3 and review for midterm</td></tr>
<tr><td>Week 8</td><td colspan="2">Midterm exam: *GIS for Homeland Security* chapters 1 and 5, *GIS Tutorial for Homeland Security* chapters 1–3</td></tr>
<tr><td>Week 9</td><td>*GIS for Homeland Security*, chapter 4</td><td>Chapter 4   Designing map layouts for homeland security planning and operations: *Prevent*</td></tr>
<tr><td>Week 10</td><td></td><td></td></tr>
<tr><td>Week 11</td><td>*GIS for Homeland Security*, chapter 2</td><td>Chapter 5   Analyzing data for homeland security planning and operations: *Protect*</td></tr>
<tr><td>Week 12</td><td></td><td></td></tr>
<tr><td>Week 13</td><td>GIS for Homeland Security, chapter 3</td><td>Chapter 6   Analyzing data for homeland security planning and operations: *Respond*</td></tr>
<tr><td>Week 14</td><td>GIS for Homeland Security, chapter 6</td><td>Chapter 7   Analyzing data for homeland security planning and operations: *Recovery*</td></tr>
<tr><td>Week 15</td><td></td><td>Conclude Chapter 7 and Review for Final Exam</td></tr>
<tr><td>Week 16</td><td colspan="2">Final exam: *GIS for Homeland Security* chapters 2, 3, 4, and 6; *GIS Tutorial for Homeland Security* chapters 4–7</td></tr>
</table>

The following table provides a sample schedule for the planning of instructional resources for a course designed with an 8-week/6 contact hours per week structure.

<table>
<tr><td colspan="3" align="center">**GIS Tutorial for Homeland Security**
Proposed timeline
8 weeks, 6 contact hours per week</td></tr>
<tr><td colspan="3">**Resources**
Kataoka, *GIS for Homeland Security*
Radke and Hanebuth, *GIS Tutorial for Homeland Security*</td></tr>
<tr><td>**Date**</td><td>**Text chapter**</td><td>***GIS Tutorial for Homeland Security* chapter**</td></tr>
<tr><td>Week 1</td><td>*GIS for Homeland Security*, chapter 1</td><td>Chapter 1   Introduction to GIS and homeland security</td></tr>
<tr><td>Week 2</td><td>*GIS for Homeland Security*, chapter 1</td><td>Chapter 2   Visualizing data for homeland security options</td></tr>
<tr><td>Week 3</td><td>*GIS for Homeland Security*, chapter 5</td><td>Chapter 3   Compiling data for homeland security planning and operations: *Prepare*</td></tr>
<tr><td>Week 4</td><td></td><td>Conclude chapter 3 and review for midterm</td></tr>
<tr><td></td><td colspan="2">Midterm Exam: *GIS for Homeland Security* chapters 1 and 5, *GIS Tutorial for Homeland Security* chapters 1–3</td></tr>
<tr><td>Week 5</td><td>*GIS for Homeland Security*, chapter 4</td><td>Chapter 4   Designing map layouts for homeland security planning and operations: *Prevent*</td></tr>
<tr><td>Week 6</td><td>*GIS for Homeland Security*, chapter 2</td><td>Chapter 5   Analyzing data for homeland security planning and operations: *Protect*</td></tr>
<tr><td>Week 7</td><td>*GIS for Homeland Security*, chapter 3</td><td>Chapter 6   Analyzing data for homeland security planning and operations: *Respond*</td></tr>
<tr><td>Week 8</td><td>*GIS for Homeland Security*, chapter 6</td><td>Chapter 7   Analyzing data for homeland security planning and operations: *Recover*</td></tr>
<tr><td></td><td colspan="2">Final exam: *GIS for Homeland Security* chapters 2, 3, 4, and 6; *GIS Tutorial for Homeland Security* chapters 4–7</td></tr>
</table>

The following table provides a sample schedule for the planning of instructional resources for a course designed with a 10-week/4.5 contact hours per week structure.

GIS Tutorial for Homeland Security		
Proposed timeline		
10 weeks, 4.5 contact hours per week		
Resources		
Kataoka, *GIS for Homeland Security*		
Radke and Hanebuth, *GIS Tutorial for Homeland Security*		
Date	**Text chapter**	***GIS Tutorial for Homeland Security* chapter**
Week 1	*GIS for Homeland Security*, chapter 1	Chapter 1 Introduction to GIS and homeland security
Week 2	*GIS for Homeland Security*, chapter 1	Chapter 2 Visualizing data for homeland security operations
Week 3	*GIS for Homeland Security*, chapter 5	Chapter 3 Compiling data for homeland security planning and operations: *Prepare*
Week 4		
Week 5		Conclude chapter 3 and review for midterm
	Midterm exam: *GIS for Homeland Security* chapters 1 and 5, *GIS Tutorial for Homeland Security* chapters 1–3	
Week 6	*GIS for Homeland Security*, chapter 4	Chapter 4 Designing map layouts for homeland security planning and operations: *Prevent*
Week 7	*GIS for Homeland Security*, chapter 2	Chapter 5 Analyzing data for homeland security planning and operations: *Protect*
Week 8	*GIS for Homeland Security*, chapter 3	Chapter 6 Analyzing data for homeland security planning and operations: *Respond*
Week 9	*GIS for Homeland Security*, chapter 6	Chapter 7 Analyzing data for homeland security planning and operations: *Recover*
Week 10		Conclude chapter 7 and review for final exam
	Final exam: *GIS for Homeland Security* chapters 2, 3, 4, and 6; *GIS Tutorial for Homeland Security* chapters 4–7	

The following table provides a sample schedule for the planning of instructional resources for a course designed with a 5-week/9 contact hours per week structure.

GIS Tutorial for Homeland Security		
Proposed timeline		
5 weeks, 9 contact hours per week		
Resources		
Kataoka, *GIS for Homeland Security*		
Radke and Hanebuth, *GIS Tutorial for Homeland Security*		
Date	**Text chapter**	**Chapter title**
Week 1	*GIS for Homeland Security*, chapter 1	Chapter 1 Introduction to GIS and homeland security Chapter 2 Visualizing data for homeland security options
Week 2	*GIS for Homeland Security*, chapter 5	Chapter 3 Compiling data for homeland security planning and operations: *Prepare* Conclude chapter 3 and review for midterm
	Midterm exam: *GIS for Homeland Security* chapters 1 and 5, *GIS Tutorial for Homeland Security* chapters 1–3	
Week 3	*GIS for Homeland Security*, chapters 4, 2	Chapter 4 Designing map layouts for homeland security planning and operations: *Prevent* Chapter 5 Analyzing data for homeland security planning and operations: *Protect*
Week 4	*GIS for Homeland Security*, chapter 3	Chapter 6 Analyzing data for homeland security planning and operations: *Respond*
Week 5	*GIS for Homeland Security*, chapter 6	Chapter 7 Analyzing data for homeland security planning and operations: *Recover* Conclude chapter 7 and review for final exam
	Final exam: *GIS for Homeland Security* chapters 2, 3, 4, and 6; *GIS Tutorial for Homeland Security* chapters 4–7	

Pedagogical guidelines and considerations for scenario 1

- This tutorial book is set up to flow in sequential order.
- The user does not have to have any previous experience with ArcGIS or any of its components prior to taking this course.
- Each chapter contains sections designated as "Your Turn." These provide the students with the opportunity to apply skills and knowledge they have just acquired.
- The PowerPoint presentations will introduce the concepts that will be covered in each chapter along with actual maps that will be created within that chapter.

Scenario 2: Professional development tool for homeland security administrators, first responders, and GIS professionals

In this scenario, *GIS Tutorial for Homeland Security* is used as a resource for a professional development environment. The sequential order of the book helps professionals with different perspectives to use GIS tools in their focus areas. It also helps them understand homeland security-related focuses important to other professions, thus encouraging a symbiotic relationship among professionals in all avenues. In a formal training environment, participants can greatly benefit from collaborative groups. In this context, trainers should determine appropriate content based on participants' level of understanding of GIS technology and career experience. In a self-study environment, the participant can have the flexibility to spend additional time focusing on chapters that have particular relevance to his/her career while also be introduced to how GIS technology is used in other focus areas of homeland security.

GIS Tutorial for Homeland Security concept map:

Chapter 3	Chapter 4	Chapter 5.1
Compiling data for homeland security planning and operations	DHS National Planning Scenario 5: Chemical attack—blister agent	DHS National Planning Scenario 12: Explosive attack—bombing using improvised explosive device

Chapter 3

MISSION AREA: *PREPARE*
Delineate DHS Urban Area

Prepare Minimum Essential Data Set (**MEDS**) database:

    Boundaries

    Transportation

    Hydrography

    Structures

    Orthoimagery

    Land cover

    Elevation

    Geographic names

Chapter 4

MISSION AREA: *PREVENT*
DHS Target Capability: Information sharing and collaboration.
DHS Task: Pre.A.5 3 Disseminate indications and warnings.

MISSION AREA: *PROTECT*
Incident-site and EOC actions to dispatch, detect, assess, predict, monitor, and sample hazard material.

MISSION AREA: *RESPOND*
Alerts, activation and notification; traffic and access control; protection of special-needs populations; resource support; requests for assistance; and public information.

MISSION AREA: *RECOVER*
Decontamination of immediate concentrated and distant spot contamination; disposal of contaminated waste; environmental testing; and public provision.

Chapter 5.1

MISSION AREA: *PREVENT*
Detection in preevent planning stages.

MISSION AREA: *PROTECT*
DHS Target Capability: Critical infrastructure protection program.
DHS Task: Pro.B1 Implement protection measures.

MISSION AREA: *RESPOND*
Alerts, activation and notification; traffic and access control; protection of special-needs populations; resource support; requests for assistance; and public information.

MISSION AREA: *RECOVER*
Decontamination, removal and disposal of debris and remains; repair and restoration of main venue and transportation center.

Chapter 5.4

DHS National Planning Scenario 6: Chemical attack—toxic industrial chemicals

MISSION AREA: *PREVENT*
Visibly increasing security and apprehension potential at the site before and during the attack.

MISSION AREA: *PROTECT*
DHS Target Capability: Citizen protection.
DHS Task: Pro.C.2.1.9 Provide public safety, develop protection plans for special-needs populations.

MISSION AREA: *RESPOND*
Alerts, activation and notification; traffic and access control; protection of special-needs populations; resource support; requests for assistance; and public information.

MISSION AREA: *RECOVER*
Decontamination of concentrated areas; disposal of contaminated wastes; environmental testing; repair of destroyed/damaged facilities; and public-information activities.

Chapter 6

DHS National Planning Scenario 10: Natural disaster—major hurricane

MISSION AREA: *PREVENT*
NHC/FEMA video teleconference forecast and impact assessments.

MISSION AREA: *PROTECT*
Review impact assessments for infrastructure and rapid resource needs; request remote sensing products; run path, size and intensity models; assess search and rescue, medical, and navigation needs.

MISSION AREA: *RESPOND*
DHS Target Capability: Search and rescue.
DHS Task: Res.B.4 Conduct search and rescue.

MISSION AREA: *RECOVER*
Prepare for secondary hazards and events; assess property and structural damage from flooding and high winds; assess and recover utility service disruptions; assess public health threat from disease and hazardous materials; and assess economic repercussions of disrupted business operations.

Chapter 7.1

DHS National Planning Scenario 9: Natural disaster—major earthquake

MISSION AREA: *PREVENT*
Event not preventable.

MISSION AREA: *PROTECT*
Retrofit critical infrastructure; conduct response training exercises for emergency operations personnel and population.

MISSION AREA: *RESPOND*
EPA/USCG manage hazardous material spills; American Red Cross delivers emergency medical treatment, shelters and food; Joint Information Center distributes instructions to the public; urban search and rescue deployed.

MISSION AREA: *RECOVER*
DHS Target Capability: Restoration of lifelines.
DHS Task: Rec.C.3 Provide energy-related support.

Chapter 7.2

DHS National Planning Scenario 9: Natural disaster—major earthquake

MISSION AREA: *PREVENT*
Event not preventable.

MISSION AREA: *PROTECT*
Retrofit critical infrastructure; conduct response training exercises for emergency operations personnel and population.

MISSION AREA: *RESPOND*
EPA/USCG manage hazardous material spills; American Red Cross delivers emergency medical treatment, shelters and food; Joint Information Center distributes instructions to the public; Urban search and rescue deployed.

MISSION AREA: *RECOVER*
DHS Target Capability: Structural damage assessment and mitigation.
DHS Task: Rec.C.2.3.1 Post-incident assessment of structures, public works, and infrastructure.

Define GIS
Define spatial data for graphic and image map layers
Review the national infrastructure for spatial data
Review the unique capabilities of GIS
Demonstrate how GIS can be used for homeland security operations
Introduce ArcGIS and its user interface

Chapter 1

Introduction to GIS and homeland security planning and operations

Geographic information systems (GIS) technology offers unique and valuable applications for policy makers, planners, and managers in many fields, including homeland security operations. GIS software and applications enable you to visualize and process data in ways never before possible. The purpose of this book is to provide you with hands-on experience using the premier GIS software package, ArcGIS Desktop, in the context of homeland security operations. You need not have any previous experience using GIS.

The next section describes GIS, its inputs and special capabilities, followed by a discussion of homeland security operations and GIS applications. There is also a preview of the upcoming chapters in this book, and an introduction to ArcGIS in a short tutorial exercise.

What is GIS?

GIS is a multidisciplinary software system that engages geographers, computer scientists, social scientists, planners, engineers, and others. Consequently, it has been defined from several different perspectives.[1] A preferred definition emphasizes GIS as an information system: GIS is a system for input, storage, processing, and retrieval of spatial data. Except for the additional word "spatial," this is a standard definition for an information system. Spatial components include a digital map infrastructure, GIS software with unique functionality based on location, and new mapping applications for organizations of all kinds. A definition and discussion of these distinctive aspects of GIS follows.

Spatial data

Spatial data is information about the locations and shapes of geographic features, in the form of either vector or raster data. Graphic maps, also known as vector maps, are created with layers that have features drawn using points, lines, and polygons. Geographic objects have a variety of shapes, but all of them can be represented as one of three geometric forms—a polygon, a line, or a point. A polygon is a closed area with a boundary consisting of connected straight lines. For example, figure 1.1 is a map with one polygon map layer (state boundaries), a line layer of roads (USA Interstates), and a point layer (cities with populations greater than 100,000). Fill color is used here within the state polygons to show the total population per state as of the 2000 census.

Associated with individual point, line, or polygon features are data records that provide identifying and descriptive data attributes. For example, in figure 1.1 the labels for the names of states and cities come from tables of attribute records associated with each map layer.

Figure 1.1

Raster maps are aerial photographs, satellite images, or images created with software that are stored in standard digital image formats, such as tagged image file format (TIFF) or Joint Photographic Experts Group (JPEG). An image file is a rectangular array, or raster, of tiny square pixels. Each pixel has a single value and solid color, and corresponds to a small, square area on the ground, from six inches to three feet on a side for high-resolution images. (Pixels are also referred to as cells.) Accompanying the image files are world files or headers that provide georeferencing data, including the upper left pixel's location coordinates and the width of each pixel in ground units. The world file or header provides the data needed for the GIS software to assemble individual raster datasets into larger areas and overlay them with aligned vector datasets.

Viewed on a computer screen or a paper map, a raster map can provide a detailed backdrop of physical features. Figure 1.2, an aerial photograph overlaid with vector map layers, shows locations around a bridge plaza, a noted piece of critical transportation infrastructure where security and surveillance operations are upgraded if a homeland security threat increases.

Figure 1.2

Map layers have geographic coordinates, projections, and scale. Geographic coordinates for the nearly spherical world are measured in polar coordinates, angles of rotation in degrees, minutes, and seconds, or decimal degrees. The (0,0) origin is generally taken as the intersection of the equator with the prime meridian (great circle) passing through the poles and Greenwich, England. Longitude is measured to the east and west of the origin for up to 180° in each direction. Latitude is measured north and south for up to 90° in each direction.

The world is not quite a sphere because the poles are slightly flattened and the equator is slightly bulged out. The world's surface is better modeled by a spheroid, which has elliptical cross sections with two radii, instead of the one radius of a sphere. The mathematical representation of the world as a spheroid is called a datum; for example, two datums commonly used for North America are NAD 1927 and NAD 1983. If you use the same projection, but with two different datums, then each corresponding map will have small but noticeable differences in coordinates.

A point, line, or polygon feature on the surface of the world is on a 3D spheroid, whereas features on a paper map or computer screen are on flat surfaces. The mathematical transformation of a world feature to a flat map is called a projection. There are many projections, some of which you will use in exercises throughout this tutorial. Each projection has its own rectangular coordinate system with a (0,0) origin conveniently located so that coordinates generally are positive and have distance units, usually feet or meters.

Necessarily, all projections cause distortions of direction, shape, area, and lengths in some combination. So-called conformal projections preserve shape at the expense of distorting area. Some examples are the Mercator and Lambert conic projections. Equal area projections are the opposite of conformal projections: they preserve area while distorting shape. Examples are the cylindrical and Albers equal area projections.[2]

Map scale is often stated as a unit-less, representative fraction; for example, 1:24,000 is a map scale where 1 inch on the map represents 24,000 inches on the ground and any distance units can be substituted for inches. Small-scale maps have a vantage point far above the earth and large-scale maps are zoomed in to relatively small areas. Distortions are considerable for small-scale maps but negligible for large-scale maps relative to policy, planning, and research applications.

GIS maps are composites of overlaying map layers. For large-scale maps such as in figure 1.2, the bottom layer can be a raster map with one or more vector layers on top, placed in order so that smaller or more important features are on top and not covered up by larger contextual features. Small-scale maps, such as figure 1.1, often consist of all vector map layers. Each vector layer consists of a homogeneous type of feature: points, lines, or polygons.

Digital geospatial data infrastructure

GIS is perhaps the only information technology that requires a major digital infrastructure: namely, a collection of standards, codes, and data designed, built, and maintained by government. We also refer to the map layers of the infrastructure as basemaps. Vendors provide valuable enhancements to the digital map infrastructure, but for the most part, it is a public good financed by tax dollars. Without this infrastructure, GIS would not be a viable technology.

The National Spatial Data Infrastructure (NSDI), developed by the Federal Geographic Data Committee (FGDC), " ... encompasses policies, standards, and procedures for organizations to cooperatively produce and share geographic data."[3] The U.S. Department of Homeland Security (DHS) has adapted and integrated these policies, standards and procedures into the Geospatial Data Model (GDM) specific to the needs of homeland security planners, analysts, and operations personnel. The GDM is a standards-based, logical data model to be used for collection, discovery, storage, and sharing of homeland security geospatial data. The model supports development of the department's services-based geospatial architecture, and serves as an extract, transform, and load (ETL) template for content aggregation.[4]

FGDC, in conjunction with the National Geospatial-Intelligence Agency (NGA), and the U.S. Geological Survey (USGS), developed minimum essential datasets compliant with the GDM to conduct missions in support of homeland defense and security. These datasets, which will be fully explored in later chapters of this tutorial, draw from a variety of sources to compile geospatial information about places throughout the country. A primary source of basic geospatial data is the TIGER/Line maps provided by the U.S. Census Bureau. These maps are available by states and counties for many classes of layers. These classes and examples of each follow:

- *Political layers*—states, counties, county subdivisions (towns and cities), and voting districts
- *Statistical layers*—census tracts, block groups, and blocks
- *Administrative layers*—ZIP Codes and school districts
- *Physical layers*—highways, streets, rivers, streams, lakes, railroads, and landmarks

You can download TIGER/Line map layers in GIS-ready formats at no cost from a variety of Web sites, including the U.S. Census Bureau's (www.census.gov) and ESRI's (www.esri.com). Steps for doing so are found in chapter 3.

Corresponding to TIGER/Line maps of statistical boundaries is census demographic data tabulated by census areas such as tracts and blocks. Data from the decennial census is available at no cost from www.census.gov/geo/www. Steps for downloading census data and preparing it for GIS use are in chapter 5.

The USGS is the "largest water, earth, and biological science and civilian mapping agency ... [and it] collects, monitors, analyzes, and provides scientific understanding about natural resource conditions, issues, and problems."[5] Among its products useful for homeland security applications are its 1:24,000-, 1:25,000-, and 1:63,360-scale topographic maps, known as digital raster graphic (DRG) maps in scanned image format, and its digital orthophoto quarter quadrangle (DOQQ) aerial photographs (such as used in figure 1.2). Full national coverage of the most recent DOQQs and 90 percent of DRG maps are available at little to no cost from the National Map Seamless Server (NMSS) at seamless.usgs.gov. Full instructions on viewing and downloading data from this portal are included in chapter 3.

Regional and local governments provide many of the large-scale map layers in the United States, and in this data most features are smaller than a city block. Included are deeded land parcels and corresponding real property data files on land parcels, structures, owners, building roof footprints, utility networks, and pavement digitized from aerial photographs. Often you can obtain such map layers and data for nominal prices from local governments, though they have become increasingly difficult to obtain due to security and privacy issues that have arisen since the attacks on September 11, 2001.

Unique capabilities of GIS

Historically, maps were made for reference purposes. Examples of reference maps are street maps, atlases, and the USGS topographic maps. It wasn't until GIS, however, that analytic mapping became widely possible. For analytic mapping, an analyst collects and compiles map layers for the problem at hand, builds a database, and then uses GIS functionality to provide information for understanding or solving a problem. Before GIS, analytic mapping was limited to a few kinds of organizations, such as city planning departments. Analysts did not have digital map layers, so they made hard-copy drawings on acetate sheets that could be overlaid and switched in and out, for example, to show before-and-after maps for a new facility such as a baseball stadium. With GIS, however, anyone can easily add, subtract, turn on and off, and modify map layers in an analytic map composition. This capacity has led to a revolution in geography and an entirely new tool for organizations of all kinds.

As figures 1.1 and 1.2 show, maps use symbols that are defined in map legends. Graphical elements of symbols include color fill, pattern, and boundaries for polygons; width, color, and type (solid, dashed, etc.) for lines; and shape, color, and outline for points. A GIS analyst does not apply symbols individually to features, but applies and renders a layer at a time based on attribute values associated with geographic features.

For example, given a code attribute for roads with values primary, secondary, and local, a GIS analyst can choose a yellow, bold line for primary roads; a red, thin line for secondary roads; and an orange dashed line for local roads. Those three steps render all roads in a map layer with different line symbols determined by the road type.

Similarly, the color-shaded state map layer in figure 1.1 is based on an attribute that provides the 2000 population by state. A map that uses color fill in polygons for coding is called a choropleth map. In this case it shows a natural-breaks numeric scale, rendered using a green color scale. The darker the shade of green, the higher the interval of the numeric scale. By making selections and setting parameters, the GIS analyst accomplishes all of this coding and rendering with a simple graphical user interface.

Most organizations generate or collect data that includes street addresses, ZIP Codes, or other georeferences. GIS is able to spatially enable such data, that is, add geographic coordinates or make data records joinable to boundary maps. Geocoding, also known as address matching, uses street addresses as input and assigns point coordinates to address records on or adjacent to street centerlines, such as in the TIGER/Line street maps. Geocoding uses a sophisticated program that has built-in intelligence—similar to a postal delivery person's when getting your mail to you—that can interpret misspellings, variations in abbreviations, and so on.

Policy, planning, and research activities often require data aggregated over space and time, rather than individual points. For example, in a study to determine the number of on-site incidents within proximity to emergency service locations, it may be desirable to aggregate address data to counts per census tract or ZIP Code boundaries. GIS has the unique capacity to determine the areas in which points lie, using a spatial join or overlay function, and this enables the analyst to count points or summarize their attributes (e.g., using sums or averages) by area.

How does GIS support homeland security planning and operations?

This book has a sampling of homeland security GIS applications. It is prepared as a guide to support the series of homeland security initiatives driving the federal government's efforts to safeguard the country, primarily the National Preparedness Goal, National Planning Scenarios, Target Capabilities List, and Universal Task List. These initiatives combine to provide an "all-threats/all-hazards" guide to local, state, tribal, and federal governments in their efforts to build collaborative and integrated capabilities to secure the homeland.

The **National Preparedness Goal** (the Goal), a result of Homeland Security Presidential Directive 8, is designed "to achieve and sustain risk-based target levels of capability to prevent, protect against, respond to, and recover from major events, and to minimize their impact on lives, property, and the economy, through systematic and prioritized efforts by Federal, State, local and Tribal entities, their private and nongovernmental partners, and the general public."[6] The Homeland Security Grant Program (HSGP) is a primary funding mechanism for building and sustaining national preparedness capabilities.[7] The *FY2007 Homeland Security Grant Program* document is a guide to securing funding available to government agencies through the Department of Homeland Security, which makes available approximately $1.7 billion in grant funding to build capabilities that enhance homeland security. This GIS tutorial parallels the target initiative of the Goal, by providing a sampling of GIS applications that directly support mission area efforts to *prevent*, *protect* against, *respond* to, and *recover* from major events that impact our national security.

The **National Planning Scenarios** is a set of "15 all-hazards-planning scenarios for use in national, Federal, State, and local homeland security preparedness activities. These scenarios are designed to be the foundational structure for the development of national preparedness standards from which homeland security capabilities can be measured because they represent threats of hazards of national significance with high consequences."[8] This tutorial employs a sample of these scenarios within the context of the Goal to demonstrate how GIS is applied to the range of possible hazards and attacks, including natural disasters, and biological, chemical radiological, and explosives attack. It is not intended to encompass all potential threats or hazards, but rather a selection of security situations that may be enhanced by the application of GIS tools and analysis.

The **Target Capabilities List** (TCL)[9] is a set of 36 capabilities to support capabilities-based planning, which is defined as, "planning, under uncertainty, to provide capabilities suitable for a wide range of threats and hazards while working within an economic framework that necessitates prioritization and choice."[10] The underlying reasoning that drives the application of this TCL is that if a community has met the set of capabilities targeted to address a particular threat or hazard, then it is prepared to prevent, respond to, or recover from that event. Each exercise within this tutorial employs a particular national scenario, and applies selected GIS techniques to demonstrate how they support a particular target capability identified as critical to planning for that threat or hazard. For example, in chapter 6, GIS is used to demonstrate how the target capability of *Search and Rescue* can be met and enhanced as response personnel geocode an address database to locate missing persons in the aftermath of a natural disaster such as a major hurricane.

The **Universal Task List** (UTL)[11] is the basis for defining the capabilities found in the TCL that are needed to perform the full range of tasks required to prevent, protect against, respond to, and recover from incidents of national significance.[12] Approximately 300 tasks have been identified as *critical* to the success of a homeland security mission. Each exercise in this tutorial uses a selected set of these defined critical tasks within the context of a target capability to meet the challenge of a particular homeland security mission. For example, chapter 5 demonstrates how the implementation of protective measures around critical infrastructure can be aided and enhanced by GIS tools and techniques such as 3D visualization and line-of-sight analysis.

The Goal specifically identifies the essential role that geospatial analysis plays in a successful homeland security mission:

> "DHS Office of Grants and Training recognizes the important contribution that geospatial information and technology plays in strengthening our Nation's security posture. Federal, State and local organizations have increasingly incorporated geospatial information and technologies as tools for use in emergency management and homeland security applications. Geospatial data and systems improve the overall capability and information technology applications and systems to enhance public security and emergency preparedness and efficient response to all-hazards including both natural and man-made disasters."[13]

The *FY 2007 Homeland Security Grant Program Supplemental Resource: Geospatial Guidance* document focuses exclusively on GIS with respect to homeland security operations. DHS has developed a standards-based geospatial data model for GIS systems built and used at all levels of government for collection, discovery, storage and sharing of geospatial data.[14] Compliance with this model will ensure an open, interoperable and shareable system, which is a critical imperative at a time of security crisis.

In support of this geospatial model, federal, state, and local governments worked together to create **Minimum Essential Data Sets** (MEDS) over urbanized and large areas, and for national critical infrastructure to fulfill the Joint Forces Command Common Relevant Operating Picture. The MEDS provide the geospatial foundation necessary for the homeland security community to carry out the key national homeland security strategy objectives, as outlined by the While House on July 16, 2002: (1) preventing terrorist attacks within the United States; (2) reducing the nation's vulnerability to terrorism; and (3) minimizing damage, while speeding recovery from natural or terrorist-caused disasters.[15]

The MEDS data layers include:

- Orthoimagery
- Elevation
- Hydrography
- Transportation
- Boundaries
- Structures
- Land cover
- Geographic names

All of the exercises in this tutorial use selected MEDS data layers to build, process, and analyze geospatial data to support capabilities-based planning for homeland security. The scale of these MEDS data layers is determined by the geographic extent at which the planning effort is focused. For urban areas, the MEDS should have the currency and positional accuracy qualities typically sought by local governments. For larger areas at a smaller scale (states or groups of states), these datasets should have the positional accuracy qualities of the USGS primary topographic map series.[16]

Beyond the content of the MEDS data itself, the most fundamental issue with respect to homeland security planning and operations is the collaboration and delivery of data and support systems necessary to make critical decisions during a crisis. The use of geospatial data in a homeland security operation can be integrated at the desktop level, multiuser server configuration, multiserver federated enterprise platform, or a geo-Web-distributed server environment managed and accessed via the Internet. The scenarios studied in this tutorial are designed to be managed at any one of these levels of planning and operations. The GIS tools and analysis introduced here can be applied regardless of the platform or environment through which the data is delivered and managed though a desktop environment does contain more "out of the box" capability than a multiserver federated enterprise system or a fused geo-Web environment where much of the geospatial analysis capability requires a great deal of interoperability, collaboration, and applications development to fully function as needed.

Below is an overview of each chapter in this tutorial, providing the context of how GIS supports homeland security planning and operations.

Chapter 1: Introducing GIS and homeland security planning and operations
This first chapter introduces the basic parameters and functionality of ArcGIS within the context of homeland security operations. The primary goal of the DHS is to coordinate the creation of homeland security operations from state to state, and urban area to urban area, within an integrated and standardized framework. As noted earlier, the Department of Homeland Security recognizes the critical importance of geospatial data analysis in any effort to protect the nation's communities from hazards and attacks. The chapters with exercises that support and strengthen this geospatial component of homeland security operations are summarized below.

Chapter 2: Visualizing data for homeland security planning and operations
Chapter 2 introduces the basic ArcMap components and explores them using the extensive suite of navigation and viewing tools. You will apply these tools to basic point, polyline, and polygon layers that are relevant to homeland security operations. This chapter also explores the link between the map and attribute tables to reveal the capability to manage geospatial data in both a map and tabular format. Attributes of U.S. cities with population greater than 100,000 and U.S. interstate highways are selected, identified, and found to demonstrate how these tools aid the user in viewing and navigating a map display.

Chapter 3: Preparing data for homeland security planning and operations
Chapter 3 encompasses the preparation of essential geospatial data layers in a comprehensive dataset critical to the safety and security of the nation. Having such a geospatial dataset in place at the time of a catastrophic event aids and assists emergency operations personnel in protecting, responding to, and recovering from such an event. In this chapter, you will explore and build the components of MEDS, as defined by DHS. The MEDS include data from various Internet portals, and are prepared in a range of formats, including vector shapefiles, raster datasets, tables, and geodatabases. Each of the databases will be downloaded, and processed for inclusion into the MEDS database using a full suite of ArcMap geoprocessing

tools, to include select, buffer, merge, import, export, clip, and adding x,y coordinate data as a data layer. In later exercises, you will use portions of this data to address the prevention, protection, response, and recovery of communities in crisis scenarios.

Chapter 4: Designing map layouts for homeland security planning and operations: Prevent
Chapter 4 extends the basic ArcMap skills learned in earlier exercises to preparing and producing reports and map layouts. The target capability of *information sharing and collaboration* is met by performing the task of *disseminating indications and warnings* within the context of National Planning Scenario 5: Chemical attack—blister agent. Map layouts are composed of a suite of elements that define the map information presented. Prepared and custom map templates help create effective warning reports and map layouts that can be printed on paper or exported in a variety of digital formats for electronic distribution. Within the *Prevent* Mission Area of homeland security planning, this set of skills is essential to disseminating critical information that may actually prevent an attack.

Chapter 5: Analyzing data for homeland security planning and operations: Protect
Chapter 5 applies the suite of ArcMap tools and geospatial analysis to the *Protect* Mission Area of homeland security planning and operations. This chapter focuses on two national planning scenarios to demonstrate how GIS tools and analysis protect the nation's people and places.

The target capability of *critical infrastructure protection (CIP)* is met by performing the task of *implementing protection measures* within the context of National Planning Scenario 12: Explosive attack—bombing using improvised explosive devices. Buffering, selection, and intersect tools are employed to identify critical infrastructure within an area of a potential threat. Additional 3D spatial tools, such as viewshed and line of sight, are applied to optimize positioning of surveillance sites around targeted facilities.

The target capability of *citizen protection: evacuation and/or in-place protection* is met by performing the task of *providing public safety-develop protection plans for special needs populations* within the context of National Planning Scenario 6: Chemical attack—toxic industrial chemicals. Incorporating census data into the GIS, homeland security planners and operations personnel can effectively locate and identify how many people, including those with special needs, may be affected by an emergency event. This assists in allocating necessary relief and evacuation resources before an event to ensure the safety of those affected. Spatial selection tools are also employed to identify the location of assembly points for evacuation outside an affected area.

Chapter 6: Analyzing data for homeland security planning and operations: Respond
Chapter 6 brings you to the aftermath of Hurricane Katrina as emergency response workers converge on the Gulf Coast to locate and rescue missing persons. The target capability of *search and rescue* is met by performing the task of *conducting search and rescue* within the context of National Planning Scenario 10: Natural disaster—major hurricane.

The exercises in this chapter include geocoding a set of street addresses of the last known location of missing persons; and then preparing a suitability map to identify available sites for helicopter landing zones in the affected areas near the geocoded locations of missing persons. The geoprocessing steps are then programmed into a model to automate the sequence of operations involved in the analysis so that it can be executed each time inundation conditions change along the coast. When the analysis is complete, the U.S. National Grid is draped over the map layout to provide a nationally defined coordinate system for spatial referencing, mapping, and reporting, as required by DHS.

Chapter 7: Analyzing data for homeland security planning and operations: **Recover**
In the final chapter of this tutorial, GIS tools and analysis are applied to the *Recover* Mission Area of homeland security planning and operations. The target capability of *restoration of lifelines* is met by performing the task of *providing energy-related support*. Recovery includes immediately restoring essential services to a community after a major disaster or event. Using National Planning Scenario 9, focusing on a major earthquake in an urban area, the USGS ShakeMaps are studied to determine San Francisco Bay Area locations expected to experience the greatest impact. Within these areas, GIS is used to identify segments of high-pressure gas-transmission lines most susceptible to rupture during a major catastrophic earthquake. You will also trace power disruptions sustained along a fault line and identify knock-on effects of these outages on infrastructure, such as other public utilities, hospitals, churches, and schools that are critical to a recovery effort. Reports and maps of parcels and services experiencing outages are prepared for distribution. This information is essential to emergency operations personnel as they coordinate efforts with utility workers to restore service to impacted residents.

Once lifeline restoration is under way, focus shifts to the assessment of damage to private property incurred by individual property owners. The target capability of *structural damage assessment and mitigation* is met by performing the task of *post-incident assessment of structures, public works, and infrastructure*. Industry-standard damage assessment tools are integrated into the GIS to enable rapid and efficient evaluation of properties located in the greatest hazard potential zones. Hyperlinking photos enhances GIS functionality by providing a visual record of assessed properties, and reports and maps are generated to track the restoration effort for planning and operations purposes.

Chapter 1: Data dictionary

Layer	Type	Layer Description	Attribute	Attribute Description
USA_Cities.shp	Shapefile	USA Cities points		
USA_Interstates.shp	Shapefile	USA Interstate highway polylines		
USA_States.shp	Shapefile	USA State polygons	POP2000	2000 Census state population

Introduction to ArcGIS components

ArcGIS Desktop consists of ArcView, ArcEditor, and ArcInfo software. All three of the ArcGIS Desktop products look and work the same, though they differ in how much they can do.

This book is designed for use with ArcView 9.3 software. ArcView, the most popular member of this collection and most widely used GIS package in the world, is a full-featured GIS software application for visualizing, managing, creating, and analyzing geographic data. ArcEditor adds more GIS editing tools to ArcView. ArcInfo is the most comprehensive GIS package, adding advanced data conversion and geoprocessing capabilities to ArcEditor. While this book was written for ArcView, you can use it with ArcEditor and ArcInfo as well.

Also available as part of ArcGIS is the free ArcReader mapping application, which allows users to view, explore, and print maps. Finally, ArcGIS has numerous extensions to ArcView, ArcEditor, and ArcInfo. Some major extensions include ArcGIS 3D Analyst for 3D rendering of surfaces, ArcGIS Network Analyst for routing and other street network applications, and ArcGIS Spatial Analyst for generating and working with raster maps.

ArcView consists of two application programs, ArcCatalog and ArcMap. ArcCatalog is a utility program that has file browsing, data importing and converting, and file maintenance functions (such as create, copy, and delete)—all with special features for GIS source data. For managing GIS source data, you will use ArcCatalog instead of the Windows utilities My Computer or Windows Explorer.

GIS analysts use ArcMap to compose a map from basemap layers, and then can carry out many kinds of analyses and produce several GIS outputs. A map composition is saved in a map document file with a name chosen by the user and the .mxd file extension. For example, in this chapter you will open GISHS_C1E1.mxd, a map document already created for use in this tutorial.

A map document stores pointers (paths) to map layers, data tables, and other data sources for use in a map composition, but does not store a copy of any data source. Consequently, map layers can be stored anywhere on your computer, local area network, or even on an Internet server, and be part of your map document. In this tutorial, you will use data sources available from the data DVD accompanying the book, as well as data prepared by you from datasets downloaded from various Web portals. Both types of data will be stored on your desktop computer's hard drive for use in the tutorial exercises.

Installing ArcView and the homeland security tutorial data DVD

This book includes a DVD with the 180-day trial version of ArcView. See appendix B for instructions on installing ArcView and the data DVD also accompanying the book. You must successfully install ArcView and the data to complete the exercises in this tutorial.

Exercise 1.1
Introducing the ArcGIS user interface

The following steps will acquaint you with the functionality and user interfaces of ArcMap and ArcCatalog. You will start by using ArcCatalog to browse the data sources used in figure 1.1, and then examine the completed project itself. You will learn how to build, modify, and query data in the remaining chapters.

In the exercises that follow, you need to be at your computer to carry out the numbered steps. Screenshots accompanying the steps show you important dialog boxes and output. Occasionally we have added "Your Turn" exercises after a series of steps. It's critical that you do these exercises, which do not take much time, to start internalizing the processes covered.

Launch ArcCatalog

ArcCatalog is the program you use to organize and manage various geospatial datasets and documents that you use in ArcMap. This program allows you to connect to your data source locations, browse through your workspaces, examine or explore the data, manage data, tables and metadata, and search for data and maps.

1   **From the Windows taskbar, click Start, All Programs, ArcGIS, ArcCatalog.**

Depending on how ArcGIS and ArcMap have been installed, you may have a different navigation menu or a name other than ArcGIS.

2   **Navigate to the \ESRIPress\GISTHS folder.**

All of the tutorial materials are contained in this folder. Each chapter contains a labeled subfolder where the map documents and tutorial data are located (e.g. \ESRIPress\GISTHS \GISTHS_C1). All of the work that you will do in this tutorial will be saved in the \ESRIPress\GISTHS\MYGISTHS_Work folder, and named using your first and last initials to uniquely identify all of your work.

3   **Click the small plus sign next to the GISTHS folder to expand it to see its contents.**   ⊞ 📁 GISTHS

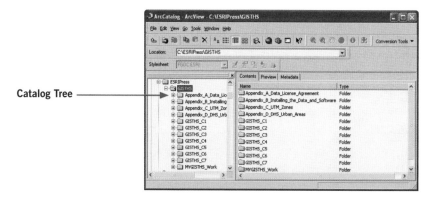

Catalog Tree ———→

The left panel of ArcCatalog is called the Catalog Tree. It is used to navigate to the data on your computer or network server, much like Windows Explorer.

4   Expand the **GISTHS_C1** folder and click on the folder name.

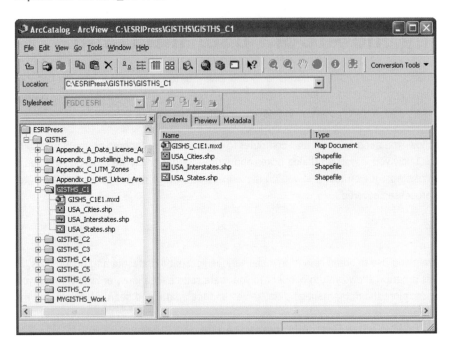

The right panel, called the Catalog Display, contains three tabs: Contents, Preview, and Metadata. When you choose the Contents tab, the datasets in the current folder are listed. The datasets currently listed represent spatial data, and the icon next to each file name indicates what type of geometry the data is built with: point, line, or polygon.

5   In the Catalog Display, click USA_States.shp, then click the Preview tab at the top of the right panel.

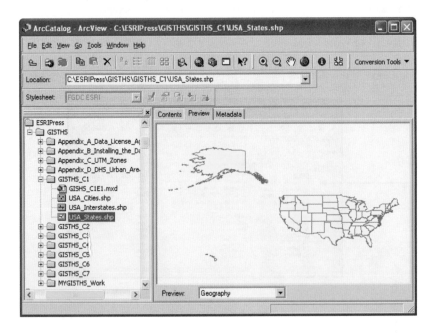

Previewing data this way allows you to get a quick glimpse of the data without actually loading it into a map. You can also use this tab to preview the contents of a table.

6   At the bottom of the Catalog Display, click the Preview drop-down arrow and click Table. Use the horizontal scroll bar to view the attribute fields in the table.

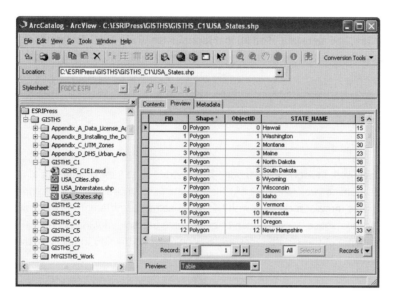

Each record in the table corresponds to one of the state polygons you previewed in the previous step, and as you can see, there are quite a few attributes stored for each state, most of which are demographic. For example, by reading across the table, you could identify that the State of Washington is in the Pacific subregion and has a population of 5,894,121 in 2000. The State_Name attribute was used to label the states in figure 1.1.

7   Click the Metadata tab at the top of the Catalog Display, and click the Spatial tab in the resulting display. (If the Spatial tab does not display, make sure the Stylesheet drop-down list is set to FGDC ESRI.)

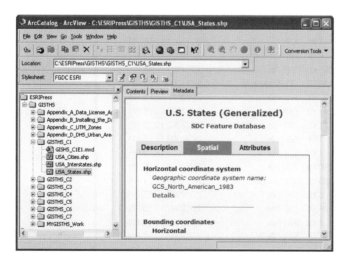

Metadata describes data; it is data about data. For example, you can see that States has a geographic coordinate system (latitude and longitude) with certain bounding coordinates for the rectangle framing the map layer. Also available are descriptions of the data and a list of attributes with their data types, under the Description and Attributes buttons.

EXERCISES

8   Click the Contents tab in the Catalog Display.

TURN
YOUR

Explore the contents, preview, and metadata of additional layers in the GISTHS_C1 folder. When finished, click the contractor button with the minus sign to the left of the GISTHS_C1 folder icon in the catalog tree.

Review data layer types

You will do most GIS file maintenance work in ArcCatalog, though it is instructive to view GIS files in a conventional browser. Next you will examine two common ESRI file formats used in GIS: a shapefile and a file geodatabase. A shapefile map layer has three or more files with the same name but different file extensions, all stored in the same folder. A file geodatabase is a collection of geographic datasets of various types stored as relational database tables. Both of these data types are very common in the GIS industry, with the geodatabase being a more modern form.

1   From the Windows taskbar, click Start, My Computer. (The path to My Computer may differ depending on which operating system you are using.)

2   Browse to \ESRIPress\GISTHS\GISTHS_C1.

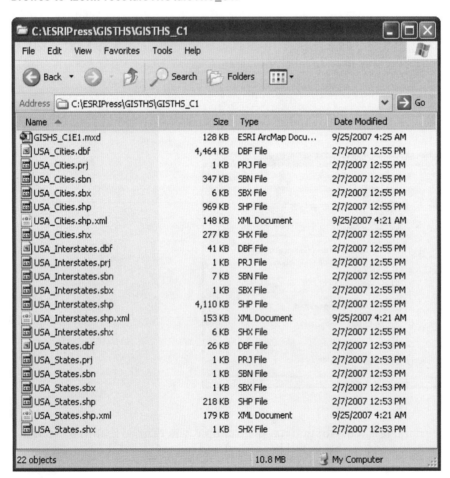

EXERCISES

Let's review the USA_Cities shapefile, which actually consists of seven files, all with the name USA_Cities.

- The USA_Cities file with .shp extension has the feature's geometry and coordinates. In this case, each record has a point and an x,y location. For line and polygon layers, each shape record has coordinates of a line segment or a polygon.

- The USA_Cities file with the .dbf extension has the feature attribute table in dBASE format. This file can be opened and edited in Microsoft Excel and Access, but such work must be done carefully and without deleting or adding records or changing the order of rows. This could result in corrupted data. The relationship between the .shp and .dbf files of a shapefile depends on one-to-one physical arrangement of records in both files.

- The .sbx, .sbn, and .shx files contain indexes for speeding up searches and queries.

- The .prj file is a simple text file that has the map projection parameters of the layer.

- Finally, the .shp.xml file contains the layer's metadata and can be opened in a Web browser for reading.

3   Click Up One Level 🗁 in My Computer.

4   Double-click **GISTHS_C6** to view the contents of this folder.

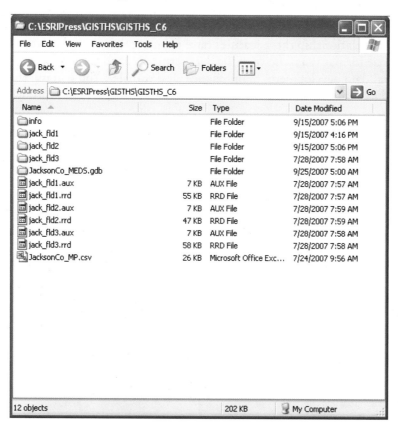

The folder JacksonCo_MEDS.gdb, is a geodatabase of features stored in a relational database.

5   Double-click **JacksonCo_MEDS.gdb** to view the file contents of this geodatabase.

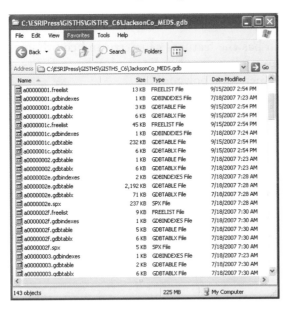

This set of files defines the features in the database. The features are stored in point, line, and polygon format here, as in the shapefile, but this database can also contain additional information that defines the relationship these features have to each other. It can also contain raster files. This geodatabase format is a more versatile data structure, and is the preferred format for future geospatial data development, maintenance, and transfer. You will use the conventional shapefile and raster image formats, as well as the geodatabase format throughout this tutorial.

6   Close the file explorer window

7   Return to ArcCatalog and view this geodatabase in the catalog display.

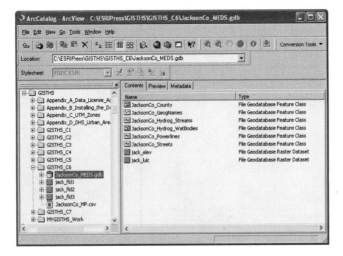

The geodatabase is labeled with a cylinder icon ⬚ in the Catalog Tree, and each feature is labeled with a feature type icon, either point, line, polygon, or raster in the Catalog Display. While it is possible to view all of these data formats using your operating systems file manager, never open any of them up in anything other than ArcCatalog or ArcMap to avoid corrupting the layers.

Launch ArcMap and open a new empty map document

ArcMap is where you display and explore the datasets for your study area, where you assign symbols, and where you create map layouts for printing or publication. ArcMap is also the application you use to create and edit datasets.

1   From the Windows taskbar, click Start, All Programs, ArcGIS, ArcMap.

Depending on how ArcGIS and ArcMap have been installed, you may have a different navigation menu or a name other than ArcGIS. Alternately, if ArcCatalog is still open, you can also open ArcMap by clicking the ArcMap button 🔍 on the ArcCatalog standard toolbar.

2   The first time you start ArcMap, the Startup dialog box appears. Click the "A new empty map" radio button.

3   Click OK.

A new untitled empty ArcMap document opens.

EXERCISES

Review ArcMap menus, buttons, and tools

Menu bar
Standard toolbar
Tools toolbar

Map display

Table of contents

Status bar

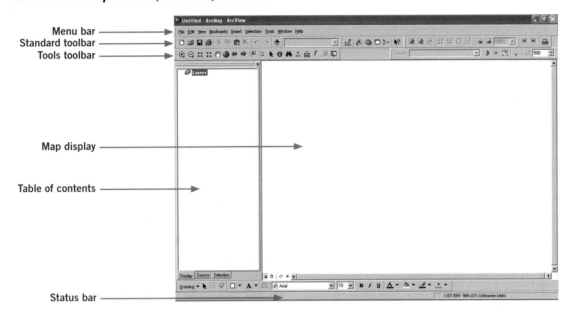

Let's review the major components of the ArcMap application window.

- The Menu Bar has some items common to most window application packages, plus some unique to GIS.

- The Standard Toolbar most typically appears at the top of the ArcMap application window and is used for map printing, creating a new map, opening an existing map, saving your map, and starting related ArcGIS applications.

- The Tools Toolbar has frequently used mapping tools and can be undocked by dragging and dropping it to the desired location.

- The Map Display is where the datasets loaded into the map are drawn.

- The Table of Contents lists all the data in the map document and allows you to toggle their visibility and access their properties.

- The Status Bar shows the map coordinates of the cursor location in the map display.

Open an existing map document

You will now open a map document that is already created. It contains the map of the United States with the U.S. interstate highways, and cities with greater than 100,000 population that you viewed earlier in this exercise.

1   From the Standard toolbar, click the Open button.

2   Browse to **\ESRIPress\GISTHS\GISTHS_C1** and select **GISTHS_C1E1.mxd**.

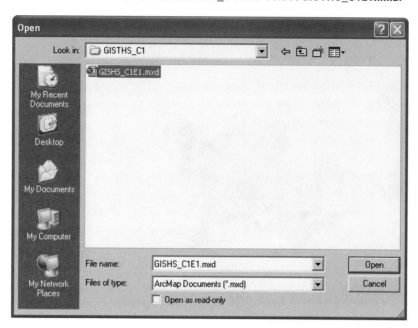

3   **Click OK.**

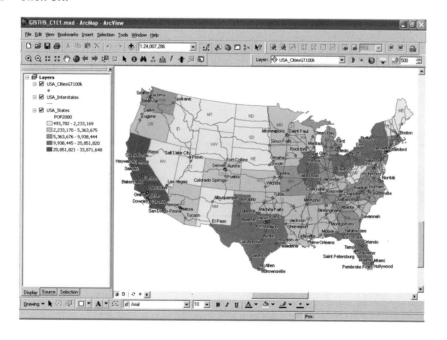

4   In Table of Contents, click the checkbox next to the USA_Interstates data layer to uncheck it.

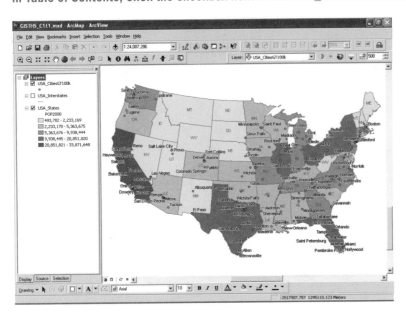

Notice how this data layer disappears from the map display.

5   Close the ArcMap document without saving changes.

This ArcMap document has already been prepared for you for use in this exercise. In the next chapter you will use the same data layers to build a map from an empty untitled map document and learn how to navigate around the map and define the properties of the data layers.

YOUR TURN

Toggle the other data layers on and off to see how they are displayed on the map.

Summary

In this chapter, you were introduced to the basic principles of geographic information systems, and how data used in a GIS can be managed and viewed to enhance geospatial analysis of homeland security scenarios. The Department of Homeland Security directives and initiatives were reviewed to ascertain how geospatial analysis can be used to help emergency operations planners and personnel prevent, protect, respond, and recover from events that threaten our national security.

This introductory chapter acquainted you with the primary components of ArcGIS. ArcCatalog was reviewed to demonstrate how to view and manage the different data layer formats to include shapefile, raster, and geodatabase formats. You were introduced to the ArcMap interface, including the various menu and button bars, as well as the map display and table of contents components. You opened a preexisting ArcMap map document showing point, line, and polygon data layers symbolized to best display the data.

As you move on in this tutorial, you will discover that GIS is a fascinating and valuable information technology. It enables spatial processing and visualization of data in ways never before possible. At the most basic level, GIS makes it easy to quickly compose and render maps from base layers. You can easily turn map layers on and off to study spatial patterns and correlations to enhance situational awareness. On a more advanced level, as explored in later chapters, you can use GIS to perform complex analysis to determine how many people are impacted by a major disaster, or what critical facilities are located in proximity to a threatened site.

Notes

1. Clarke, K.C. 2003. *Getting Started With Geographic Information Systems*. 4th Edition. Prentice Hall. Upper Saddle River. pp. 2–6.

2. Ibid., pp. 42–44.

3. www.fgdc.gov

4. http://www.fgdc.gov/fgdc-news/geo-data-model

5. www.usgs.gov/aboutusgs

6. *FY 2006 Homeland Security Grant Program Guidance and Application Kit*, p. 1

7. *FY 2007 Homeland Security Grant Program: Program Guidance and Application Kit*, U.S. Department of Homeland Security, Office of Grants and Training, January 5, 2007, p. 1

8. *National Planning Scenarios*, Version 20.1 DRAFT, p. ii.

9. *Target Capabilities List: Version 1.1*, U.S. Department of Homeland Security, Office of State and Local Government Coordination and Preparedness, May 23, 2005.

10. *FY 2006 Homeland Security Grant Program*, December 2, 2005, p. 12

11. *Universal Task List: Version 2.1*, Department of Homeland Security, Office of State and Local Government Coordination and Preparedness, May 23, 2005.

12. Ibid., p. 2

13. *FY 2007 Homeland Security Grant Program Supplemental Resource: Geospatial Guidance*, January 2007, p. 1.

14. http://www.fgdc.gov/participation/working-groups-subcommittees/hswg/subgroups/info-content-sg/documents/DHS-GDM-v1.1.pdf

15. *FY 2007 Homeland Security Grant Program Supplemental Resource: Geospatial Guidance*, January 2007, p. 3.

16. *FY 2007 Homeland Security Grant Program Supplemental Resource: Geospatial Guidance*, January 2007, p. 4.

Learn the basic map components
Introduce basic ArcMap navigation and viewing tools
Apply basic tools to point, polyline, and polygon layers
Understand the link between the map and attribute tables
Manipulate attributes to select, identify, and find features

Chapter 2

Visualizing data for homeland security planning and operations

In this chapter, you are introduced to the basic functionality of ArcMap that allows you to navigate the map display and manipulate data layers. These tools include learning about the map components, viewing layer properties and attribute tables, zooming and panning around the map display, creating spatial bookmarks, using the magnifier and overview windows, measuring distance, and selecting, identifying, and finding features.

Three layers of spatial data—cities, roads, and states—are examined to familiarize you with these ArcMap tools. The Cities layer is a point feature that contains all cities in the United States with a population exceeding 100,000 people. These cities are believed to be at the greatest risk for homeland security threats and are targeted to receive the bulk of federal funding through the Urban Areas Security Initiative (UASI) program that develops plans and systems to safeguard their people and infrastructure.[1]

The Roads layer is a polyline feature of the interstate highway system, one of the most vulnerable pieces of critical infrastructure in our nation. This National System of Interstate and Defense Highways, created during President Dwight D. Eisenhower's administration, not only facilitates movements of goods and services from coast to coast but helps mobilize military and defense operations in time of emergency. Today, most of our interstate commerce is moved along this road network every day, carrying commodities essential to our communities' survival. A homeland security incident or natural hazard compromising this system could have immediate and considerable effect on our day-to-day survival.

The States layer is a polygon feature that outlines the geographic extent of each state in the country. It serves as a basemap layer for the map document.

Learning how to apply the basic viewing and query tools to these data layers at the outset will help you navigate through subsequent chapters that delve more deeply into analysis of homeland security data and scenarios.

Chapter 2: Data dictionary

Layer	Type	Layer Description	Attribute	Description
USA_Cities	Shapefile	USA Cities points		
USA_CitiesGT100k	Shapefile	USA Cities with greater than 100k population	NAME	City name
			POP2000	2000 Census state population
			STATUS	State/county legislative designation
USA_Counties	Shapefile	USA County polygons		
USA_Interstates	Shapefile	USA Interstate highway polylines	CLASS	Road classification
USA_States	Shapefile	USA State polygons		

Exercise 2.1
Manipulate map layers in a map document

In the first part of this chapter, you will learn the basics of the ArcMap software package. You will begin by opening an existing map document and then learn how to add and manipulate map layers.

Open an existing map document

1   Browse to the \ESRIPress\GISTHS\GISTHS_C2 folder.

2   Click the GISHS_C2E1.mxd map document icon, and click Open.

The resultant map of U.S. cities and states has two map layers already included and symbolized. The first layer is a point feature of major U.S. cities with population greater than 100,000. The second layer is a polygon feature of U.S. states.

EXERCISES

Add a layer

You can add more map layers to your map for more detailed analysis. For example, adding major interstate highways to this map shows the connectivity of major U.S. urban centers via the national transportation system. This data was saved as an ArcMap shapefile, which contains polyline features for the U.S. interstate roadways across the country.

1   Click the Add Data button.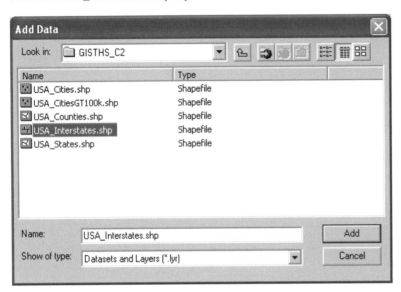

2   In the Add Data dialog box, browse to **\GISTHS_C2**.

3   Click the USA_Interstates.shp layer.

Name	Type
USA_Cities.shp	Shapefile
USA_CitiesGT100k.shp	Shapefile
USA_Counties.shp	Shapefile
USA_Interstates.shp	Shapefile
USA_States.shp	Shapefile

Look in: GISTHS_C2

Name: USA_Interstates.shp

Show of type: Datasets and Layers (*.lyr)

4   Click Add.

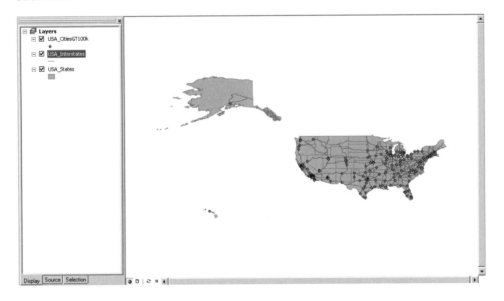

ArcMap adds this data layer to the map display, and chooses a random color that can be changed later.

EXERCISES

Save a map document

Now that you have added a new layer to the map document, you can save it with a new name in a different folder. This map document will now become your own copy that you can use to add and change data layers as needed.

1   From the ArcMap main menu, click File, Save As.

2   Browse to **\ESRIPress\GISTHS\MYGISTHS_Work** and add your first and last initials and an underscore to the beginning of the filename to rename it (e.g. FL_GISHS_C2E1.mxd).

3   Click Save.

Change a layer's display order

Changing a layer's display order is important because features may be covered up by other features in your map. ArcMap draws layers from the bottom up, from points to lines to polygons, but there may be occasions where smaller features may be hidden by larger features that overlay them.

1   Click and hold down the left mouse button on the USA_CitiesGT100k layer in the Table of Contents.

2   Drag the USA_CitiesGT100k layer to the bottom of the Table of Contents and release.

Because the Cities layer is now drawn first, its points cannot be seen. They are covered by the states and interstates layers drawn later.

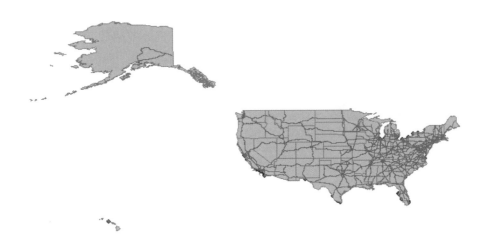

3   Click and hold down the left mouse button on the USA_CitiesGT100k layer.

4   Drag the USA_CitiesGT100k layer back to the top of the Table of Contents and release.

Because the cities layer is now drawn last, its points can now be seen again.

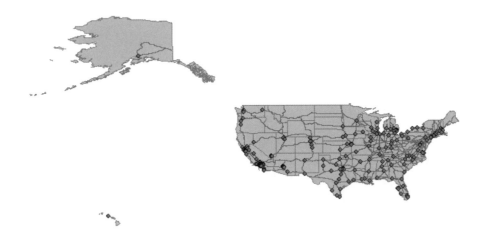

YOUR **TURN**

> Drag the USA_States layer to the top of the Table of Contents and observe what happens. Then drag it back to the bottom of the list. Do the same with the USA_Interstates layer to become familiar with how layers are best positioned and listed in the Table of Contents.

View and change data frame properties

Now that we have a map document with layers added to show the location of large urban centers and interstate highways, we can explore the properties of the map components. The data frame is the primary component of the map where layers are stored, defined, and displayed.

1   From the Table of Contents, right-click Layers and select Properties.

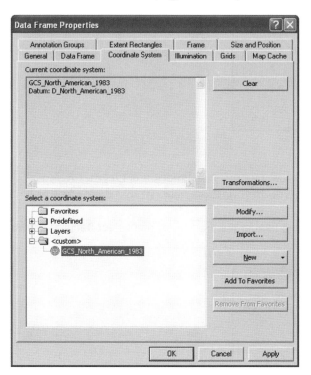

The Data Frame Properties dialog box appears and shows a set of 10 tabs that define the data frame. The Coordinate System tab is the first one to show when the dialog box is opened. This data frame is set in the GCS_North_American_1983 coordinate system. In this case, it is set by the coordinate system of the first data layer added to the data frame. We will preset and change the data frame coordinate system later in the tutorial.

EXERCISES

2   **Click the General tab.**

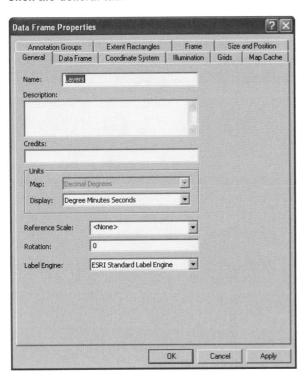

The name of this data frame is Layers. The display units designation is Degrees Minutes Seconds. These features can be changed by modifying these fields in this dialog box.

3   **Type U.S. Cities and Interstates in the Name field of the dialog box; and select Decimal Degrees from the Display drop down list.**

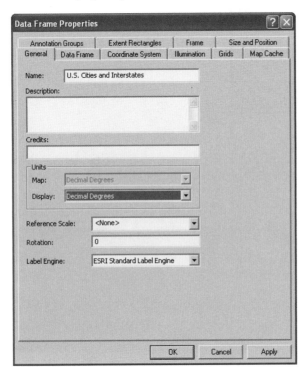

4   Click OK.

The layer is now listed in the table of contents as U.S. Cities and Interstates, and the coordinates displayed on the status bar are now shown in decimal degrees.

View map layers in a map document

A data frame contains one or more map layers. There may be times when it is not necessary to show all layers in a data frame at the same time. By turning layers on and off, other relevant data may be more visible.

1   In the Table of Contents, click the small check box to the left of the USA_CitiesGT100k layer to turn that layer off.

2   Click the small check box to the left of the USA_CitiesGT100k layer again to turn that layer back on.

YOUR TURN

Turn the other layers on and off by clicking on the check box to the left of the layer name in the Table of Contents.

EXERCISES

Exercise 2.2
View map layer properties

Each of these layers has its own set of properties that defines where it is located, how it is displayed, and how it behaves geographically.

Open layer properties and view the General layer settings

1   In the Table of Contents, right-click the USA_CitiesGT100k layer.

2   Select Properties.

3   Click the General tab to review the Layer Name, Description, Credits, and Scale Range of this layer.

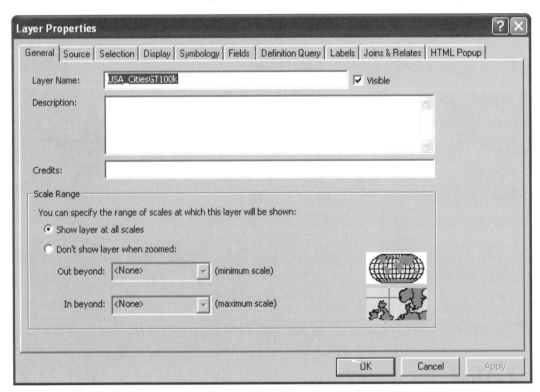

This layer is named USA_CitiesGT100k, is set to be visible on the map, has no description, and is to be shown at all scales.

Rename a layer

When ArcMap adds a layer to a map document, it uses the name of the shapefile or feature as the default name of the layer in the table of contents. You will often want to change the layer name so that it is more readable and easier to understand.

1   Type **U.S. Cities with population greater than 100,000** in the layer name field.

2   Click Apply.

3   Click OK.

Renamed layer ⟶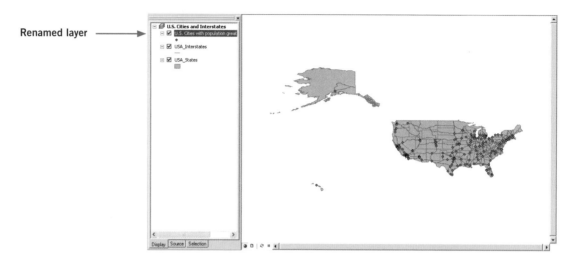

View the layer Source settings

1   Right-click the U.S. Cities with population greater than 100,000 layer again, and select Properties.

2   Click the Source tab to review the source information about this layer.

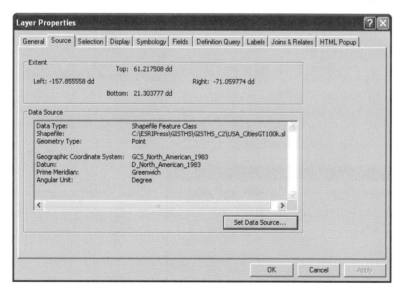

The source properties include the spatial extent of the layer (defined as a bounding box of four coordinates), and the data source of the layer. The data source information includes the data type, path, geometry, geographic coordinate system, datum, prime meridian, and angular unit. Note that while we changed the name of the layer, the original source file name of the data did not change. It is still named USA_CitiesGT100k.shp.

View the layer Selection settings

1   **Click the Selection tab to review the selection settings for this layer.**

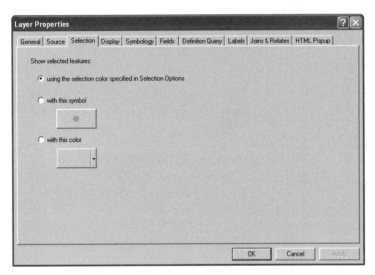

When querying this layer for a particular subset of data, such as cities with over 250,000 population, the selected data is highlighted in a bold and bright color so that it visually stands out from the rest of the map. This Selection property window enables the user to override the default selection settings of the whole map document to allow this particular layer to display differently from others when queried for selection. It is currently set to use the default Selection Options already set for the map document.

Show Map Tips in the Display settings

How a layer looks on the map can be defined by setting a few display parameters. These settings don't actually change the data itself, but only how and when it appears on the map. By modifying the display settings, the layer may become more readable and provide additional visual information about the feature.

1   **Click the Display tab to review the selection settings for this layer.**

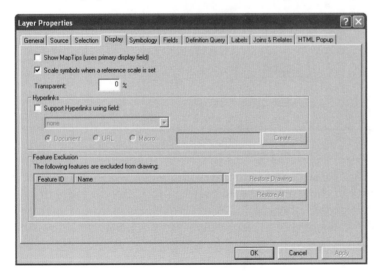

2   Click the box to the left of the Show Map Tips (uses primary display field) option.

3   Click OK.

4   From the Tools toolbar, 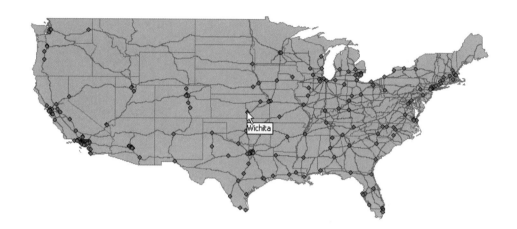 use the Select Elements tool
to scroll over the map. Stop along the way over the cities layer to see the map tips boxes as they
appear on the map showing the names of the cities.

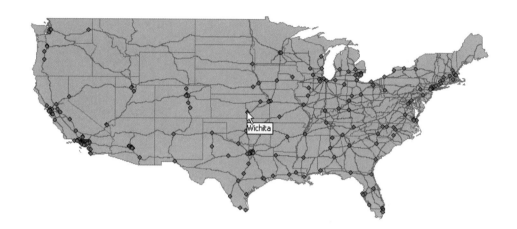

This display dialog box also allows a hyperlink to be set for this layer to enhance the map
by showing other documents when the feature is activated. A ground photo of a particular
facility located along an interstate highway may assist first responders when sent to assess an
on-site incident. Features of a layer may also be excluded from displaying to limit the extent of
data drawn to the map.

Change the Symbology settings

The most visible property of a layer is its color and shape. The symbology settings allow the user to modify
these settings to best display a feature. ArcMap assigns a set of default symbols and random colors to
features when they are added to the map document. The user almost always changes these to suit the
purpose of the map.

1   Right-click the U.S. Cities with population greater than 100,000 layer again, and select Properties.

2   Click the Symbology tab to review the settings for this layer.

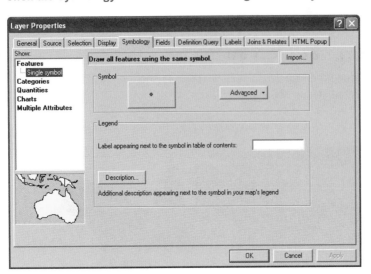

Since this layer is a point feature, ArcMap has assigned a single marker symbol in a purple color.

3   Click the symbol button to open the Symbol Selector window.

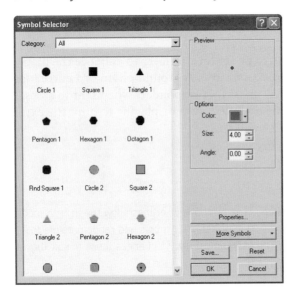

4   In the Symbol Selector window, click Circle 1.

The size of this line is preset to 18. This may be too large to show up clearly at the current map scale.

EXERCISES

5   Reduce the size of the marker to 7 and click the Color button to change the color to red.

6   Click OK.

7   Click OK again to view the new symbology of this layer.

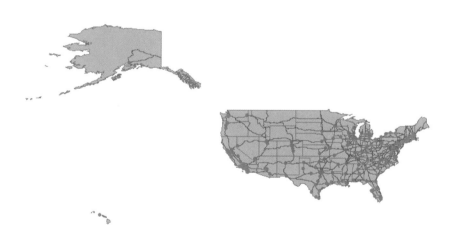

Symbology can also be modified to show different categories of data, such as quantitatively grouping cities into intervals by population thresholds. These tools will be more fully explored within the context of homeland security operations in later chapters.

YOUR TURN

Change the symbology and color of the USA_Interstates layer. You can change the size, type, and color of the line to enhance its display.

Change the Primary Display Field settings

Each map feature is accompanied by a database of fields that defines what exactly it is. The Fields tab in the Layer Properties identifies the primary display field, and lists all of the properties of the fields in the database.

1   Right-click the U.S. Cities with population greater than 100,000 layer to return to the Layer Properties.

2   Click the Fields tab.

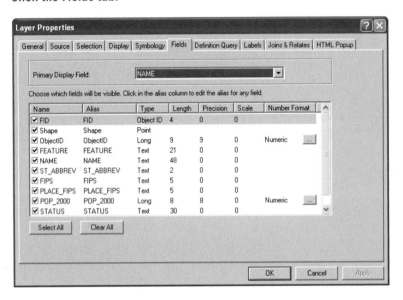

The Primary Display Field is identified as NAME. This is the same field that you saw in the Map Tips box that displayed when you scrolled over the cities on the map.

3   Click the drop down arrow to select POP_2000 as the Primary Display Field.

4   Click OK.

5   Use the Select Elements tool to scroll over the map. Stop along the cities layer to see the map tips boxes as they appear on the map showing the population of cities.

The fields settings also allow the user to set the visibility, change the alias name, and define the formatting for numeric data.

EXERCISES

Create a Definition Query

The Definition Query settings allow the user to specify if there are any particular criteria that further define this layer. For example, cities that are both state and county capitals require an added measure of security when protecting against potential threats to government infrastructure.

1   Right-click the U.S. Cities with population greater than 100,000 layer to return to the Layer Properties.

2   Click the Definition Query tab.

3   Click the Query Builder button. The Query Builder dialog box appears.

4   Scroll to the STATUS field and double click it to enter it into the SELECT box.

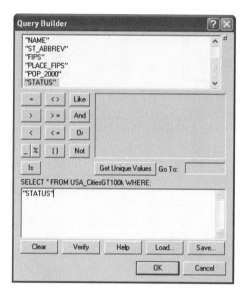

5   Click the Equals operator button.   =

6   Click the Get Unique Values button and double click "State Capital County Seat".

7   Click OK.

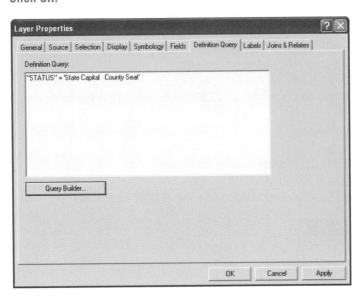

The query string "STATUS" = 'State Capital County Seat' appears in the Definition Query box.

8   Click OK.

The layer is redefined to include only those cities with population greater than 100,000 that are both state capitals and county seats.

YOUR TURN

Define the USA_Interstates layer to include only those roads that are classified as state routes ("CLASS" = "S"). When complete, return the USA_Interstates layer to include all roads.

Label a geographic feature

Labels are text items that are added to a map to annotate geographic features. The text of the label is derived from the feature's attributes, such as the name of a city.

1   Right-click the U.S. Cities with population greater than 100,000 layer to return to the Layer Properties.

2   Click the Labels tab.

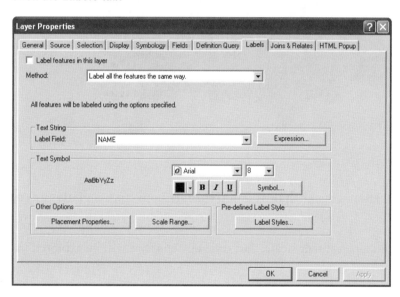

This dialog box enables the user to set the label field, and how and where the label will appear on the map.

3   Click the box to the left of the Label features in this layer option.

☑ Label features in this layer

The **NAME** field is already chosen as the Label Field.

4   Click the Text Symbol font drop-down arrow and select Lucida Console 10 Bold.

5   Click OK.

All of the cities with a population greater than 100,000 that are both state capitals and county seats are now labeled.

Other label properties include placement, scale, and style options that will be explored in later tutorials.

YOUR **TURN**

> Turn off the US Cities with population greater than 100,000 and USA_Interstates layers. Label the USA_States layer. When complete, turn off the labels for the USA_States layer.

View the layer Joins & Relates settings

1 Right-click the U.S. Cities with population greater than 100,000 layer to return to the Layer Properties.

2 Click the Joins & Relates tab.

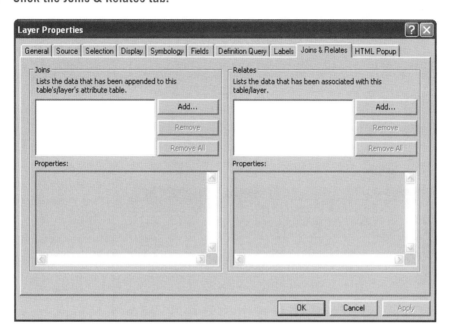

ArcMap enables the user to join and/or relate a separate set of tabular data to a geographic feature. This can be done when the two databases contain a common relate field. For example, if there is a set of tabular data that contains detailed census data for U.S. cities, it can be joined to the U.S. Cities with population greater than 100,000 layer if both databases contain the common field NAME. This feature will be explored in later exercises in this tutorial.

3 Click OK to close the Layer Properties box.

Exercise 2.3
View map layer attribute tables

All geographic features in a GIS have an attribute table associated with it. The data in the attribute table determines the geometry of the feature (point, line, or polygon), where it is located (its coordinates), and other information about the feature that define what it is. For example, in the U.S. Cities with population greater than 100,000 layer, the field NAME defines the name of the city.

Open attribute table

1   Right-click the U.S. Cities with population greater than 100,000 layer.

2   Select Open Attribute Table.

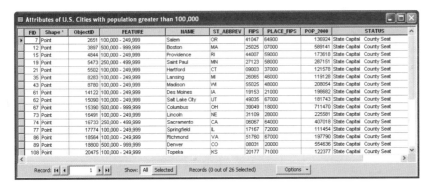

The attribute table for this layer contains 10 fields and a record, or row, for each city in the database. In this case, there are 26 cities that have greater-than-100,000 population and are both state capital and county seats. The Shape field defines the geometry and location of the feature, and the rest of the fields contain either system ID or attribute information about each of the records, or cities, in this database.

Select specific records

1   Click the record selector for Salt Lake City to select that record.

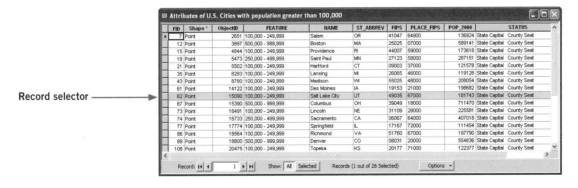

Record selector

This selected record is now highlighted in bright blue. When a record is selected in a table, it is also selected on the map, and vice versa.

EXERCISES

Show connection between layers and tables

1 Minimize the attribute table to see that the marker for Salt Lake City on the map is also selected and highlighted in bright blue.

2 Right-click the U.S. Cities with population greater than 100,000 layer.

3 Select Selection and Make This The Only Selectable Layer.

4 From the Tools toolbar, click the Select Features button and click various cities on the map. To select multiple cities, hold down the Shift key while clicking on the map.

The selected cities are now highlighted in bright blue.

5   Maximize the attribute table to see that the selected cities on the map are also selected and highlighted in bright blue in the table.

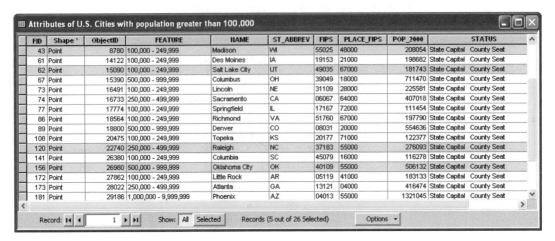

Show only selected records

1   In the attribute table, click the Selected button.

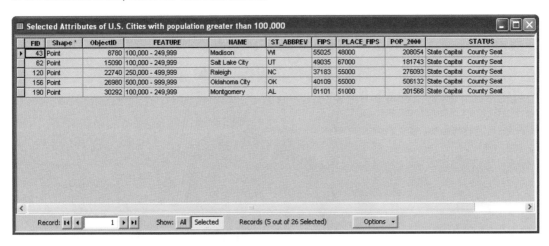

The table now shows only the selected records for the features selected in the map and table.

2   Click the All button to show all records again.

Switch selected records

1   In the attribute table, click the Options button.

2   Click Switch Selection.

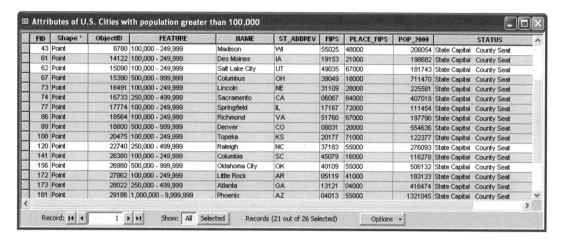

This reverses the selection by selecting all of the records that were not selected, and deselects those that were selected. This is useful when the only thing the records have in common is that they do not meet the criteria of the selected records.

Clear selected records

1   Click the Options button.

2   Select Clear Selection.

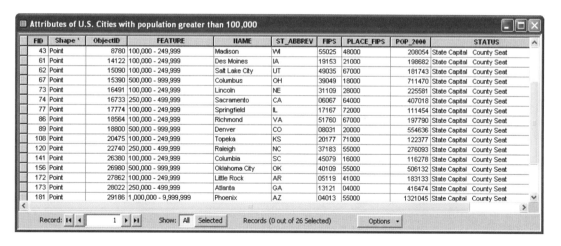

The table is returned to its original state, unselecting all records in the database.

EXERCISES

Move a field in an attribute table

Sometimes you may need to position two fields next to each other in an attribute table to better view the data. For example, in this table, the FEATURE field shows the population interval that each city falls within. Viewing this data alongside the actual population value may better reveal if a city is at the low or high end of the interval.

1   Click on the POP_2000 column heading in the attribute table.

Column heading

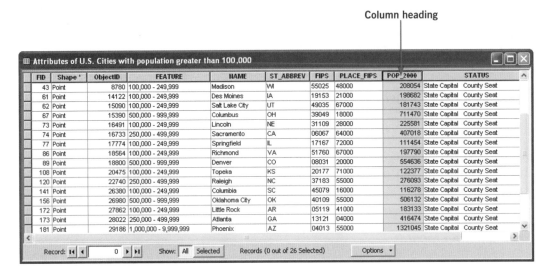

The POP_2000 field is now highlighted in bright blue.

2   Hold down the left button on the mouse and drag this column to the right side of the FEATURE field.

The POP_2000 field is now positioned next to the FEATURE field where the two datasets can be more closely compared.

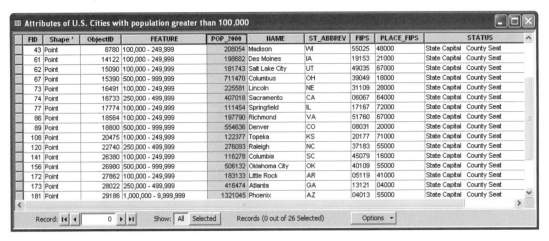

Sort a single field

You may be interested to see which of these cities that requires additional homeland security resources has the greatest population. Sorting the data in the attribute table can give you a quick view of the range of population in either an ascending or descending order.

1   Right click the POP_2000 column heading in the attribute table.

2   Select Sort Ascending.

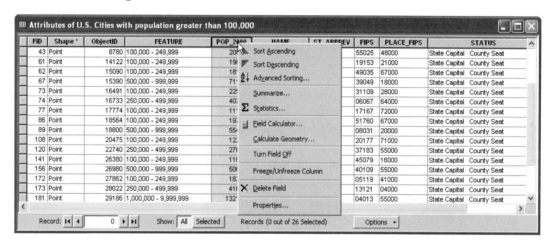

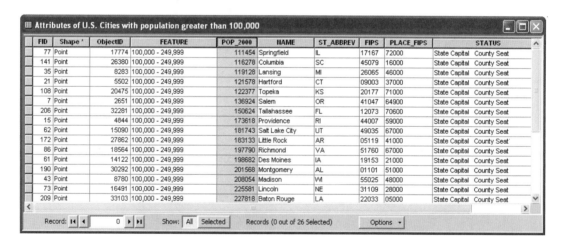

3   Scroll to the top of the attribute table.

By sorting this data in ascending order, Springfield, Illinois, is listed as the city that is both a state capital and county seat, with the lowest population of 111,454.

YOUR TURN

Sort the U.S. Cities with population greater than 100,000 layer in descending order. Which city in this database has the greatest population? Close the attribute table when complete.

EXERCISES

Exercise 2.4
Zoom and pan

Up to this point, we have viewed the map from one consistent scale. ArcMap has a suite of zoom and pan tools available on the Tools toolbar ⊕ ⊖ ▦ ▦ ✋ ⊕ ⬅ ➡ ▣ ▯ ▶ ❶ 𝘼 ⁑ₓᵧ ⬛ ⚡ ▤ ▥ that allow the user to view a map from a range of scales.

Sometimes you may need to concentrate on a particular area of a map, or want to see more clearly an area of a map that is crowded with features and labels overlapping at a scale too small to be visible.

Zoom in

1   From the Tools toolbar, click the Zoom In button.

2   Click and hold the mouse button on a point above and to the left of Madison.

3   Drag the mouse to draw a box around the four cities to include Columbus, Lansing, Madison, and Springfield.

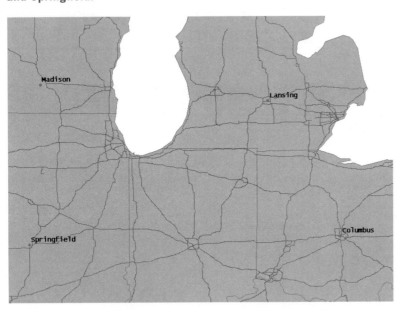

The resultant map is zoomed into these four cities. The complexity of the interstate roads is more visible at this scale, showing more clearly the interconnectedness of this transportation network.

The Zoom In tool also works when used to click on a specific point on the map.

4 Click the screen using the Zoom In tool on the city of Columbus marker.

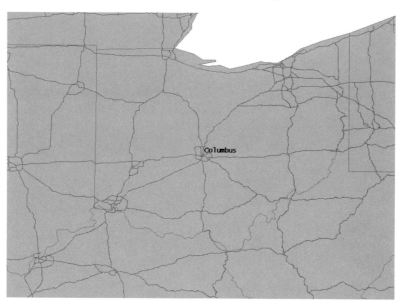

The resultant map is zoomed into and centered on the city of Columbus.

Fixed Zoom

It is also possible to use the Fixed Zoom In tool to zoom into the center of the data frame at fixed scale increments.

1 From the Tools toolbar, click the Fixed Zoom In tool five times.

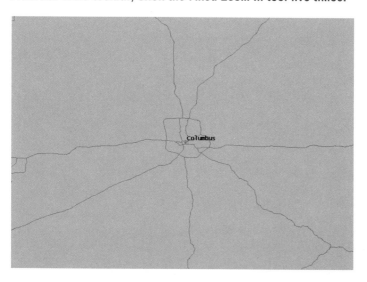

The map zooms in closer to the city of Columbus at fixed scale increments.

2 Click the Fixed Zoom Out tool ⬚ five times to return the map to the previous scale.

Pan

You can also move around the map without changing the scale by using the Pan button.

1 From the Tools toolbar, click the Pan button.

2 Click and hold the mouse button on the city of Columbus marker, and drag the mouse to the bottom edge of the data frame.

3 Release the mouse.

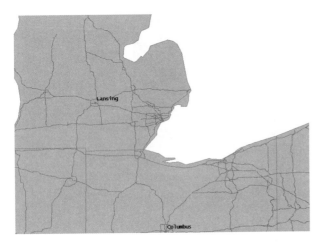

The city of Lansing is now visible, though the scale of the map remains the same. The Pan tool can be used to move left, right, up, or down in the current data frame without changing the scale.

Zoom Out

The Zoom Out tools operate in the opposite fashion as the Zoom In tools.

1 From the Tools toolbar, click the Zoom Out button.

2 Click the marker for the city of Lansing.

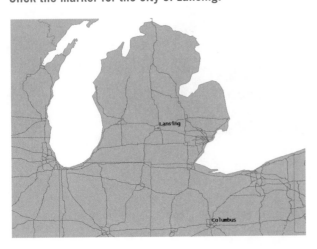

The resultant map is zoomed out from and centered upon the city of Lansing.

3   Use the Zoom Out tool to draw a wide rectangular box around the city of Lansing.

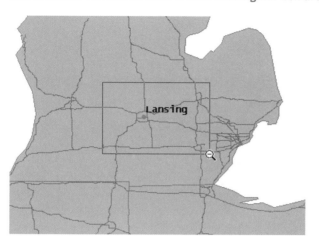

4   Release the mouse.

The resultant map is zoomed out to a smaller scale extent. By changing the size of the box drawn around a point on the map, you can change the scale extent that the map zooms out to.

5   From the Tools toolbar, click the Go Back to Previous Extent button.

6   Use the Zoom Out button to draw a tighter rectangular box around the city of Lansing.

The resultant map is zoomed way out beyond the extent of the continental United States. The smaller the box drawn using the Zoom Out tool, the further out the data frame will zoom.

Zoom to full extent

Zooming to the full extent of your map shows you all the data in your map. This tool is useful when you need to reorient yourself with the full map extent, or when you may be unsure of where you are currently viewing the data frame.

1   From the Tools toolbar, click the Full Extent button.

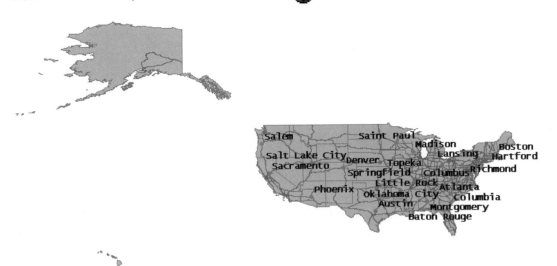

The data frame zooms to the full extent of the data.

Customize the full extent

Sometimes, though, clicking the Full Extent button zooms you out much farther than you really wanted to go. For example, since there are no cities that are both state capitals and county seats in Hawaii or Alaska, you can set the custom full extent to only the continental United States.

1   Zoom to the extent of the continental United States.

2   Right-click the data frame in the Table of Contents and click Properties.

3   Click the Data Frame tab.

4   Click Other under Extent Used By Full Extent Command.

5   Click the Specify Extent button.

6   Select Current Visible Extent.

7   Click OK twice.

8   Click the Fixed Zoom Out button three times.

9   Click the Zoom to Full Extent button.

The data frame now zooms to the full extent of only the continental United States. You can also set the custom full extent by specifying the outline of features, or by setting the coordinates of a bounding box around your preferred map extent.

Go to XY location

You can use the Go To XY command on the Tools toolbar to navigate to a particular x,y location in your map. You can specify the location by entering coordinates in the map units of your map in either Decimal Degrees, Degrees Minutes Seconds, Degrees Decimal Minutes, or using a military or U.S. National Grid system. You can pan to, zoom to, or flash the location.

1   From the Tools toolbar, click the Go to XY button.

2   Click the Units button   to set the units to Degrees Minutes Seconds

3   To zoom to Austin, Texas, enter the (x) longitude **-97 44**, and (y) latitude **30 17**.

ArcMap will automatically convert the format of the XY coordinates to the specified units.

4   Click the Zoom to tool on the Go to XY toolbar.

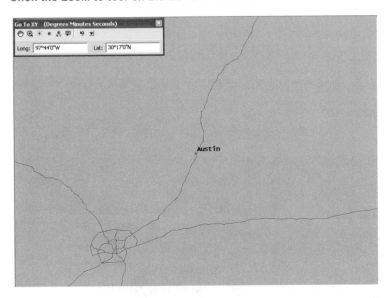

The data frame zooms to the city of Austin. You may also select to pan to the specified location, or flash the location, draw a point with or without a label, or draw a callout at the location showing its coordinates. You may also return to a previously entered location.

5   Close the Go To XY box.

YOUR TURN

Enter the coordinates of a city of your choice and add a callout box to the location. Coordinates for major cities can be found on various Web sites on the Internet.

Zoom to the extent of a layer

You can also zoom to the extent of a specific layer in the data frame.

1 Right-click the USA_States layer in the Table of Contents.

2 Select Zoom to Layer.

The map returns to the full extent of the entire United States to include all states in the data layer.

Zoom using the mouse wheel

Using the mouse wheel to zoom in and out of the data frame allows a quick and easy way to focus on a particular location on the map.

1   **Roll the mouse wheel towards you to zoom into the map.**

You can also hold down the Ctrl key to sharpen the display while you zoom.

2   **Roll the mouse wheel away from you to zoom out on the map.**

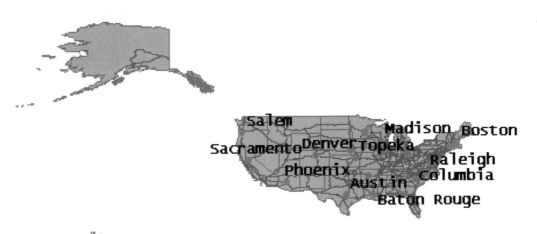

Exercise 2.5
Create spatial bookmarks

Spatial bookmarks act as place holders that allow you to repeatedly return to a particular location on your map. This is useful to move quickly to a study area or region of interest, and is helpful when making presentations.

Creating spatial bookmarks

1   Zoom to the full extent of the data frame.

2   Zoom to the city of Phoenix.

3   From the Main menu, click the Bookmarks menu.

4   Click Create and name the bookmark **Phoenix**.

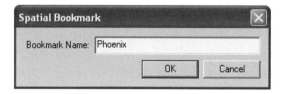

5   Click OK.

6   Zoom to the full extent of the data frame.

7   Click the Bookmarks menu and select Phoenix.

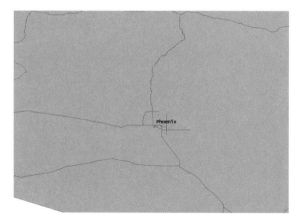

ArcMap zooms to the extent of the saved bookmark for Phoenix. This bookmark is now available every time you use this map document, which is helpful if you frequently zoom to this extent.

YOUR TURN

Create a spatial bookmark of a city of your choice.

Creating My Places

Another type of bookmark is the My Places tool. By setting preferred locations, such as addresses, x,y coordinates, features, and spatial extents, as My Places, you are placing spatial bookmarks at those map positions that can not only be repeatedly and quickly accessed in the current map document, but in any map document that you are working in.

1   Zoom to the three cities in New England that are both state capitals and county seats.

2   From the Tools menu, click My Places.

3   Click Add From and select Current Extent.

This extent is added to the My Places dialog box.

4   Zoom to the full extent of the data frame.

5   Double click the Type symbol in the My Places box to show the location of this place.

You may also add this place to your map as a graphic, labeled graphic, a call out, or a stop or a barrier along a StreetMap route.

To customize the name of this place,

6   Click on the Name field and type **New England** as the custom name for this place.

This place is always available to zoom to from this dialog box regardless of what map document is open.

7   Click Close.

EXERCISES

Exercise 2.6
Magnifier window

The Magnifier window adjusts the map display to see more or get an overview of an area on the map. This window works like a magnifying glass: as you pass the window over the map display, you see a magnified view of the location under the window. Moving the window does not affect the current map display.

Using the Magnifier window

1   From the Main menu, click Window, Magnifier.

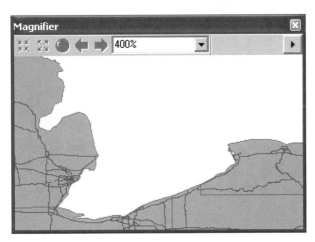

The Magnifier window appears at an unspecified location over the map display at a 400% magnification level.

2   Click the blue title bar at the top of the window and drag the magnifier over the map display until the cross hairs are positioned over Baton Rouge.

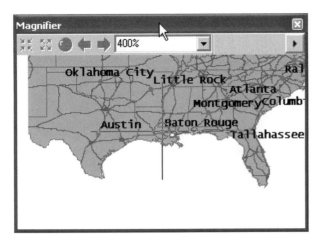

3 Release the mouse button to see the zoomed detail of the map at this location.

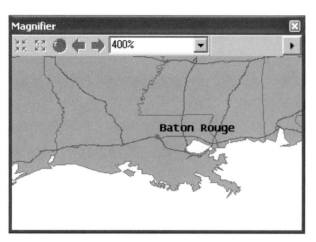

4 Click the drop-down arrow and increase the magnification to 1,000%.

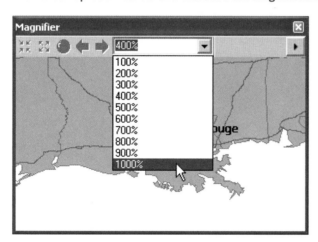

The magnifier zooms further into the selected position on the map display. You can also choose to see the magnification dynamically as the window pans over the map.

5 Click the arrow on the Magnifier window title bar.

6 Select Update While Dragging.

7 Drag the window around the map display to see the magnification as it updates.

TURN

YOUR

Change the magnification level and pan the map display over different areas. Close the Magnifier window when you are finished.

EXERCISES

Exercise 2.7
Overview window

The Overview window shows the full extent of the layers in a map while also showing the area currently zoomed to in the map display with a red box. You can move the red box to pan the map display. You can also make the red box smaller or larger to zoom the map display in or out.

Using the Overview window

1   Zoom to Oklahoma City.

2   From the Main menu, click Window, Overview.

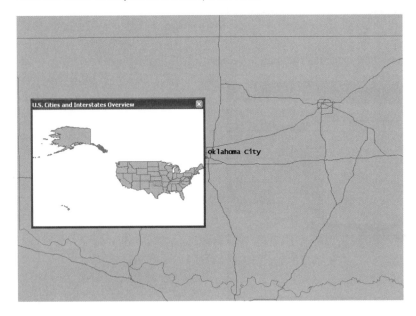

The current extent of the map display is highlighted in the Overview window with a red box.

3 Position the cursor at the center of red box and click and drag to move the Overview window to Florida.

The extent of the map display updates to reflect the changes made as you moved the Overview window to a new location. You can also change the current extent by dragging a corner of the red box in the Overview window to change the size of the box.

YOUR TURN

Drag the Overview window around the map. Drag a corner of the red box to change the size of the box to increase and decrease the map extent. Close the Overview window when you are finished.

Exercise 2.8
Viewer Window

The Viewer window allows you to create, view and dynamically update multiple extents within one map document. These views can be captured at different scales to enable a closer look at particular areas of interest while maintaining the full extent of the map in the data frame.

Using the Viewer window

1   Zoom to Saint Paul.

2   From the Main menu, click Window, Viewer.

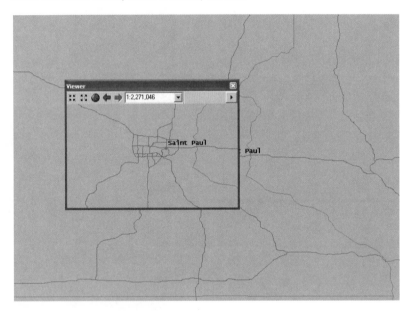

The current extent of the map display is highlighted in the Viewer window.

3   Zoom to the full extent of the data frame.

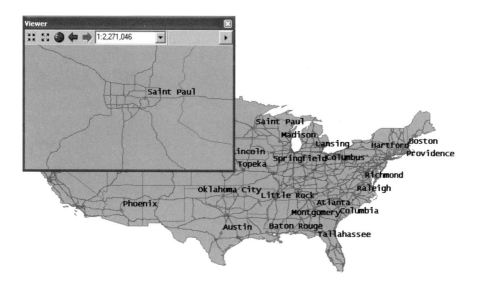

This map extent and scale is fixed in the Viewer as you change your area of interest in the data frame.

4   Change to marker color for U.S. Cities with population greater than 100,000 to a black circle symbol.

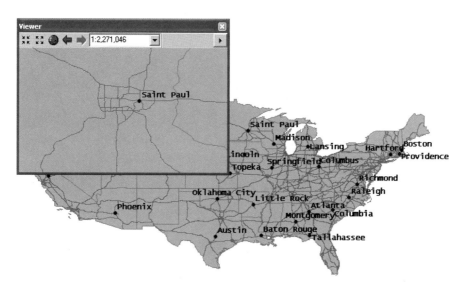

The Viewer window is also updated as you change the symbology or add new features to the map.

TURN
YOUR

Create multiple Viewer windows of areas of interest around the map. Be sure to capture them at different scales. Change the symbology of the features to see how the maps dynamically update as features are modified. Close all of the Viewer windows when you are finished.

Exercise 2.9
Measure distance

The Measure tool allows you to measure the distance of a line, or the area of a polygon. This tool is very helpful when measurements of ground distances are critical. As first responders penetrate the site of an incident, they may need to assess the range of their equipment, such as when firefighters extend hoses to the site of a fire emergency.

Using the Measure Tool

1 Zoom to Columbia, South Carolina, so you can see the route that runs from the city to the coast.

2 From the Tools toolbar, click the Measure button.

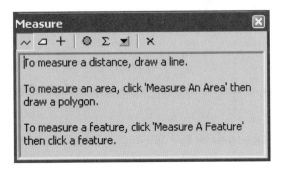

The Measure window appears. If this is the first time you are using this tool, the window has instructions on measuring the distance of line, area, or feature. It defaults to the line measurement tool. You can move this window out of the map display area if it is blocking your view of the map.

3 Click the marker for Columbia and drag the cursor down to the coast. Click the intersection before the road forks around the inlet.

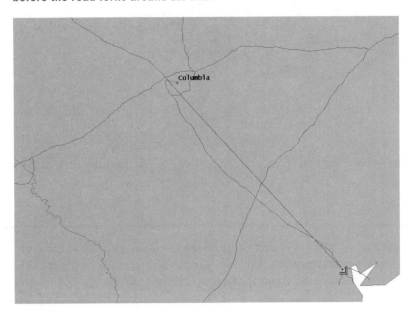

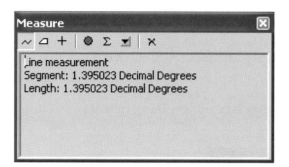

The measurements in the window show the extent of the line in decimal degrees. This unit of measure is used because the map is set in geographic coordinates.

Change measurement units

You can change the units to a more appropriate measure.

1   Click the down Choose Units button in the Measure window.

2   Click Distance, Miles.

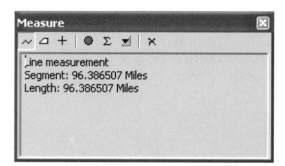

The measurements are now displayed in miles, a more useful unit.

Measuring segments and total length of a line

You can also measure the length of individual segments of a line, and at the same time show the total length of all segments.

1 **Drag the cursor over the road again and click at the intersection. Then continue to drag the cursor along the road as it leads south away from the inlet. Click at the end of the road.**

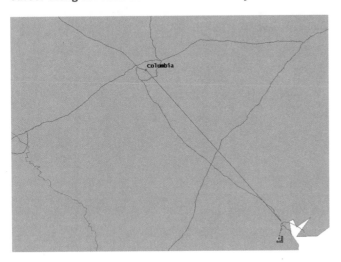

The Measure window now shows two measurements. The measurement for the segment is now 0.689871 miles, and the length of the complete line from Columbia to the coast is 105.933499 miles. Your measurements may be slightly different, as your line may not be exactly as drawn in these instructions.

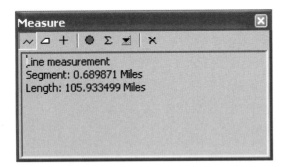

To measure the area of a polygon or feature, the data must be in a projected coordinate system. Our current data is not yet projected, but this additional feature can be used with the tutorial data specific to homeland security that is used in later chapters.

2 **Close the Measure box when complete.**

TURN

YOUR

Use the Measure tool to measure the distance of other roads on the map.

EXERCISES

Exercise 2.10
Identify features

The Identify tool displays data attributes of a feature by clicking the feature on the map. It is another way of seeing all of the fields of the attribute table interactively as you move around the map.

Identify results of a single feature

1   Zoom to the full extent of the data frame.

2   From the Tools toolbar, click the Identify button.

3   Click the marker for Sacramento.

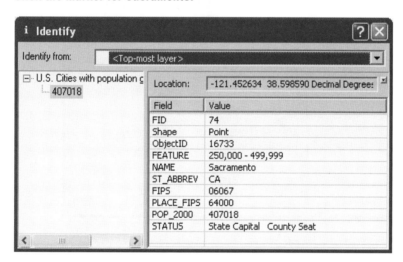

The Identify window populates with the attribute information about Sacramento. Be sure that either the <Top-most layer>, or U.S. Cities with population over 100,000 is the layer to identify from.

The layer tree shows the primary field, population (407018); the Location field shows the longitude and latitude of where you clicked (-121.452634 38.598590 Decimal Degrees; [your coordinates may be slightly different if you clicked at a nearby location]; and all of the fields and values from the attribute table are listed in the window.

EXERCISES

Identify results of multiple features

You can also view the attributes of multiple features at one time by including them in a bounding box drawn with the Identify tool.

1   Drag the Identify tool over the markers for Oklahoma City, Little Rock, Austin, and Baton Rouge.

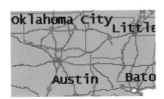

Primary field ⟶

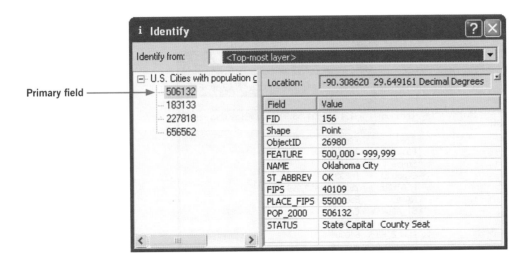

The Identify window now shows the attribute results for all four cities, listed by the primary field.

2   Click on the primary field in the layer tree to flash to the location and view the attributes of each city.

 TURN

YOUR

Use the Identify tool to view the attributes of an interstate segment, and then a state on the map. Also view the attributes of multiple features at one time.

EXERCISES

Controlling identify results

In the previous exercise you viewed the attributes of features of a single layer. The Identify tool defaults to show results for the top-most layer, but this setting can be modified to control which layer's attributes are identified for viewing.

1   Select <All Layers> from the Identify from: drop-down list.

2   Use the Identify tool to click on the marker for Sacramento.

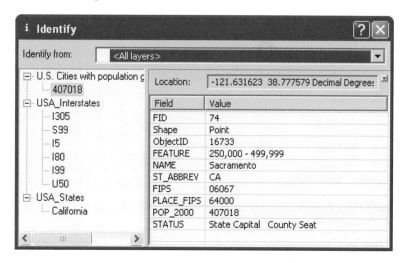

The Identify window is populated with attribute data for all of the layers that intersect at the location of the marker for Sacramento. Other options to identify from include all visible layers, selectable layers, and each individual layer in the data frame.

3   Click on the primary field in the data tree and view the attribute information of each layer.

4   Click the Field column heading to sort the fields; or click the Value column heading to sort the values.

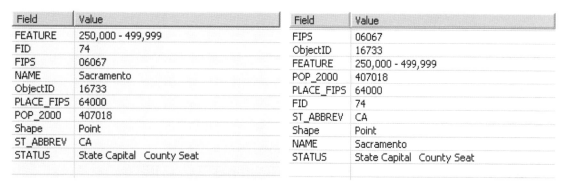

Managing the layers and attributes of the features through the Identify window enables a versatile viewing of this data in an interactive way while working directly from the map display.

5   Close the Identify box when complete.

EXERCISES

Exercise 2.11
Find features

The Find tool is used to locate features in a layer or layers based in their attribute values. Once located, you can use this tool to flash, zoom, pan, bookmark, identify, select or unselect, add the found feature to your My Places list, or find nearby places.

Find tool

1   From the Tools toolbar, click the Find button. 🔍

2   Click the Features tab in the Find dialog box, if it is not already selected.

3   In the Find: field, type **I285**.

4   From the drop-down list of the In: field, select USA_Interstates.

5   Click the radio box and search In field: ROUTE_NUM.

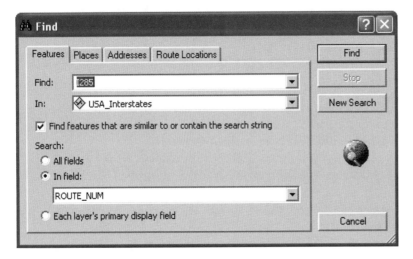

6   Click Find.

Find results

The results appear at the bottom of the Find dialog box.

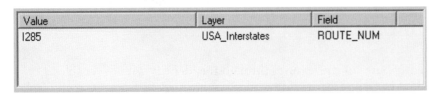

Only one interstate highway in the feature database was found with the route number I285.

1   Click the found feature in the results box.

The location will flash on the map.

2   Right click and select Zoom To.

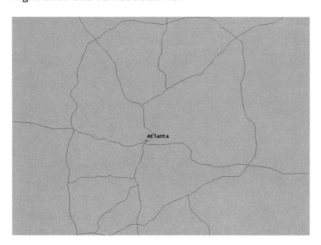

The map display zooms to this feature which is located around the city of Atlanta.

3   Right click and select Select.

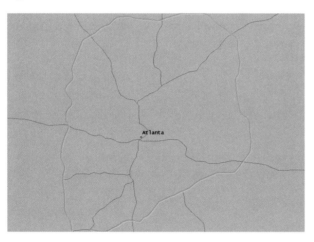

Notice that there is a road segment in this beltway that is not included in this selection.

 TURN

YOUR

Use the Identify tool to find out why this road segment is not included in the selected feature.

4   Unselect I285 and close the Find dialog box when complete.

EXERCISES

Save the project and exit ArcMap.

Earlier in this exercise, you saved this ArcMap project in the MYGISTHS_Work folder with a new name that included your initials. Now that it is already saved with this new name in a new location, you can simply save your latest changes to the same file. When you open it again, it will contain all of the changes that you made to the layers, symbology, and bookmarks throughout this exercise.

1   From the Standard button bar, click the Save button.

2   Click File, Exit to exit ArcMap.

Summary

In this chapter you learned about the basic ArcMap components and explored them using the extensive suite of navigation and viewing tools. You applied these tools to basic point, polyline, and polygon layers that are relevant to homeland security operations. The link between the map and attribute tables was also explored to reveal the capability to manage geospatial data in both a map and tabular format. Attributes of U.S. Cities with population greater than 100,000 and U.S. Interstate Highways were selected, identified, and found to demonstrate how these tools aid the user in viewing and navigating a map display.

Now that you are familiar with the basic ArcMap components, navigation, and viewing tools, you can proceed to the next chapter where you will begin to build maps relevant to homeland security planning and operations. Preparing a comprehensive dataset of essential data layers is critical to the safety and security of the nation. Having such a geospatial dataset in place during a catastrophic event helps emergency operations personnel protect, respond to, and recover from such an event.

Notes

1. "Data Count for FY2006 UASI Eligibility Determination: To identify jurisdictions for inclusion in the data count phase, all cities with a population greater than 100,000 and any city with reported threat during the past year were identified." *FY 2006 Homeland Security Grant Program*, December 2, 2005, p.77.

Chapter 3:

Compiling data for homeland security planning and operations—*Prepare*

MISSION AREA: *PREPARE*
Delineate DHS Urban Area

Prepare Minimum Essential
Data Sets **(MEDS)** database:

    Boundaries

    Transportation

    Hydrography

    Structures

    Orthoimagery

    Land cover

    Elevation

    Geographic names

3.1   Delineate DHS Urban Area
Open a new empty map document
Add boundary layer
View metadata
Select by Attributes
Create a new layer of a selected subset of features
Project a data frame
Project a data layer
Dissolve feature boundaries
Generate buffer
Adjust transparency of a data layer

3.2   Add boundary data to the MEDS map
Rename data frame
Create a group layer
Add layer in Group Layer Properties box
Select by location (intersect)
Change symbology of polygon feature
Add layer to group layer

3.3   Add transportation data to the MEDS map
Change the data frame display units
Define spatial extent of BAUA
Download transportation data from the National Map Seamless
   Server
Add a Transportation group layer
Clip layer to boundary extent
Join tables to categorize feature classes
Categorize by unique value
Export roads by Census Feature Classification Code (CFCC)
Import symbology
Show features at specified scale range
Show labels at specified scale range
Change label text symbol
Add transportation layers

3.4   Add hydrography data to the MEDS map
Download National Hydrography Geodatabase (NHD)
Add features of the hydrography geodatabase to the MEDS map

3.5   Add structures data to the MEDS map
Download landmark data from the ESRI Census TIGER/Line data
   Web Site
Merge multiple shapefiles
Define landmark shapefile coordinate system
Export landmark shapefiles to data frame coordinate system
Create new Structures Group layer
Create a layer file

3.6   Add a raster image layer to the MEDS map
Download orthoimagery from the National Map Seamless Server
Add raster orthoimagery layer to map document
Review raster layer properties
Manipulate raster image band properties

3.7   Add Land Cover layer to the MEDS map
Download land cover data from the National Map Seamless Server
Add land cover layers to map document
Export raster data layer
Assign labels to raster values and colors
Add elevation data to MEDS Map

3.8   Add a DEM layer to the MEDS map
Measure spatial extent of DEM area
Download DEM from the National Map Seamless Server

3.9   Add Geographic Names to the MEDS map
Download GNIS data from the U.S. Board on Geographic Names
   Web Site
Modify schema.ini file
Create a Geographic Names feature class from XY Table
Add GNIS shapefile to map document
Select by attributes

3.10 MEDS data management
Save all group layers as layer files

Define *prepare*, to include the creation of a comprehensive and
interoperable geospatial database
Identify the Minimum Essential Data Sets stipulated by DHS
Introduce ArcMap document management and data formats
Manipulate and query spatial data using geoprocessing tools
Explore data frame and layer projection properties
Download and integrate geospatial data from various Web portals
Explore the geodatabase model

Chapter 3

Compiling data for homeland security planning and operations: *Prepare*

To prepare is defined as "the state of having been made ready or prepared for use or action." [1] The
safety of our communities depends on our level of preparedness in three fundamental areas:

- the quality, accuracy, and currency of all types of data about a place;

- efficient and effective fusion methods to optimize data sharing and interoperability
  between agencies and jurisdictions;

- geospatial analysis using GIS to provide situational awareness at all stages of a
  homeland security operation.

During a crisis, emergency-operations personnel deploy resources based on their situational awareness of an event. The simple delivery of a common, integrated operations picture of where things are before and during an event is a major leap from the days when first responders were basically on their own in times of crisis. Integrated radio communications and common visual status maps now offer a level of situational awareness never before available across agencies and geographic jurisdictions. This allows all operations personnel to talk and see from the same vantage point during a crisis event.

Having a comprehensive geospatial database prepared, ready, available, and accessible is critical to a community's needs for prevention, preparation, response, and recovery relating to a catastrophic event. The basic task of providing access to data that displays the location of all critical facilities and population in an area is one of the most important components of a successful homeland security operation. In this chapter, you will prepare a geospatial dataset for homeland security planning and operations. The Homeland Security Infrastructure Program (HSIP), a joint effort of National Geospatial-Intelligence Agency, USGS, and the Federal Geographic Data Committee, established the development of Minimum Essential Data Sets (MEDS) to conduct their missions in support of homeland defense and security.[2] This series of minimum essential geospatial datasets is vital to a successful homeland security operation. As outlined in table 1, the MEDS data requirements are specific to Urban Area and Large Area spatial extents.

Table 1: Minimum ("no worse than") goals for resolution, accuracy, and currency [3]

Data theme	Urban areas		Large areas	
	Minimum resolution of accuracy	Minimum currency	Minimum resolution of accuracy	Minimum currency
Orthoimagery	1 foot resolution; 3 meters horizontal accuracy	2 years	1 meter resolution; 11.70 meters horizontal accuracy	5 years
Elevation	1/9 arc second (~3 meters) resolution; 0.73 meter vertical accuracy	2 years	1/3 arc second (~10 meters) (2 arc second in Alaska) resolution; vertical accuracy commensurate with contour interval of USGS primary topographic map for area	5 years
Hydrography	4.68 meters horizontal accuracy	2 years	13.90 meters horizontal accuracy; 36.69 meters horizontal accuracy for Alaska	5 years
Transportation	4.68 meters horizontal accuracy	2 years	13.90 meters horizontal accuracy; 36.69 meters horizontal accuracy for Alaska	5 years
Boundaries	4.68 meters horizontal accuracy	2 years	13.90 meters horizontal accuracy; 36.69 meters horizontal accuracy for Alaska	5 years
Structures	4.68 meters horizontal accuracy	2 years	13.90 meters horizontal accuracy; 36.69 meters horizontal accuracy for Alaska	5 years
Land cover	Should align with base maps that have the accuracies listed above.	2 years	Should align with base maps that have the accuracies listed above.	5 years
Geographic names	Same as associated feature		Same as associated feature	

The challenge is where and how to acquire these datasets. There are approximately 3,300 counties and 85,000 municipalities in the United States, most of which have some level of ongoing geospatial data-collection activities.[4] While some data may be available for certain areas, it may not be available for others. Local municipalities have over the years built extensive large-scale proprietary geospatial datasets of their communities, to include high-resolution orthoimagery, land use, parcel, utility, and transportation data. The lack of interoperability, and restrictions on data accessibility and compatibility between neighboring jurisdictions and parallel agencies during past catastrophic events such as September 11 and Hurricane Katrina, have raised the urgency for pre-event data sharing and interoperability. As a result, proprietary restrictions are gradually being lifted on local datasets to allow sharing and interoperability within and among regional associations around the country.[5]

Underpinning all efforts to promote geospatial interoperability is the National System for Geospatial-Intelligence (NSG) Standards Program. This program defines geospatial intelligence interoperability as "the ability to discover, retrieve, exploit, and exchange geospatial intelligence data and information with other systems, units, and forces, through a system of networks and services..."[6] Cooperation, data sharing, and interoperability form the backbone of the newly created Fusion Centers around the country. There are currently 38 Fusion Centers established by the DHS. These centers "blend relevant law enforcement and intelligence information analysis and coordinate security measures in order to reduce threats in local communities."[7]

More specifically,

> Analysts from the Department of Homeland Security Office of Intelligence and Analysis work side by side with state and local authorities at Fusion Centers across the country. These analysts facilitate the two-way flow of timely, accurate, and actionable information on all types of hazards.
>
> Fusion Centers
>
> * Provide critical sources of unique law-enforcement and threat information.
> * Facilitate sharing information across jurisdictions and functions.
> * Provide a conduit between men and women on the ground, protecting their local communities, and state and federal agencies.
>
> The Department will have tailored, multidisciplinary teams of intelligence and operational professionals in major Fusion Centers nationwide by the end of fiscal year 2008.[8]

Consistent with this mandate to foster data sharing and interoperability is the notion of a national geospatial response plan[9], to include a national MEDS dataset coupled with Web-enabled geospatial analytical tools, templates, and models. This model of a national geospatial response plan is embraced by the USGS National Geospatial Programs Office. The goal of this program is to align national geospatial activities and responsibilities including FGDC, Geospatial One-Stop (GOS), the National Map (TNM), and Interior Enterprise GIS into a "GIS for the Nation."[10]

The seed of this "GIS for the Nation" is found in the National Map. The National Map is the product of a consortium of federal, state, and local partners who provide geospatial data to enhance America's ability to access, integrate, and apply geospatial data at global, national, and local scales.[11] HSIP uses the National Map as its foundation to provide data for homeland security personnel to plan and operate missions nationwide. Access to the National Map is through a variety of USGS Web portals, including the National Map Viewer[12] where users view the full array of geospatial data layers making up the National Map; and the National Map Seamless Server (NMSS) where the National Map data layers are downloaded via file transfer protocol (FTP) to a local server or desktop computer for use in a GIS.

While the intent is there, and the framework is evolving to develop a fully functional GIS for the nation, it is still in its infancy at the time of this writing. While the structure of the National Map data layers closely parallels those of the MEDS, they are not all fully available and accessible for all locations. The National Map is a dynamic repository of data, and is continuously augmented with new geospatial datasets as they are developed and made available to the public. Hence, the National Map is a good place to start to build the Minimum Essential Data Sets for a successful homeland security operation. Supplemental data from local level agencies is almost always necessary at this early stage of integrated geospatial data systems. In this chapter, you will gather the MEDS data primarily from the National Map. Data from other sources will supplement the MEDS data when necessary to include data not stipulated in the MEDS or not available via the National Map.

It is important to note in this chapter that many opportunities and configurations deliver geospatial intelligence and decision support systems to homeland security operations personnel and planners. As reviewed briefly in chapter 1, geospatial data in a homeland security operation can be integrated at the desktop level, multiuser server configuration, multiserver federated enterprise platform, or a geo-Web distributed server environment managed and accessed via the Internet. The scenarios studied in this tutorial are designed to be managed at any one of these levels of planning and operations, from the centralized preparation stages to field response and recovery efforts. The GIS tools and analysis introduced in later chapters can be applied regardless of the platform or environment through which the data is delivered and managed, though a desktop environment does contain more "out of the box" capability than a multiserver federated enterprise system or a fused geo-Web environment where much of the geospatial analysis capability requires a great deal of interoperability, collaboration, and applications development to fully function as needed.

Chapter 3: Data dictionary

Layer	Type	Layer Description	Attribute	Description
cfcc.dbf	dBase Table	Census Feature Classification Database	DESCRIPTIO	Description of FCC code
			FCC	Feature Classification Code
BAUA_Airports	Shapefile	Bay Area Airport points		
BAUA_Landpt	Shapefile	TIGER Landmark points	CFCC	Census Feature Classification Code
BAUA_Landpt	Layer file	TIGER Landmark points	CFCC	Census Feature Classification Code
BAUA_Railroads	Shapefile	Bay Area Railroad polylines		
USA_Counties	Shapefile	USA County polygons	NAME	County name
USA_Places	Shapefile	USA Designated Populated Places	ST	State
			NAME	Place name
USA_States	Shapefile	USA State polygons		US State polygons
BAUA_BTSRoads\30835236.shp	Shapefile	Bay Area Bureau of Transportation	NAME	Road name
		Statistics Roads	FCC	Feature Classification Code
BAUA_DEM	Raster Dataset	Bay Area NED Digital Elevation Model		
BAUA_GeogNames: CA_DECI.txt	Text File	Geographic Names Information System	County	County name
			Class	Feature Classification
			Feature_Na	Feature Name
BAUA_LandCover\05567535	Raster Dataset	National Land Cover Data 2001 (NLCD)	VALUE	Land Cover Classification Code
			Primary_1	Latitude
			Primary_2	Longitude
BAUA_Landmarks\tgr06001lpt.shp	Shapefile	TIGER Landmark Points	CFCC	Census Feature Classification Code
BAUA_Landmarks\tgr06013lpt.shp	Shapefile	TIGER Landmark Points	CFCC	Census Feature Classification Code
BAUA_Landmarks\tgr06041lpt.shp	Shapefile	TIGER Landmark Points	CFCC	Census Feature Classification Code
BAUA_Landmarks\tgr06075lpt.shp	Shapefile	TIGER Landmark Points	CFCC	Census Feature Classification Code
BAUA_Landmarks\tgr06081lpt.shp	Shapefile	TIGER Landmark Points	CFCC	Census Feature Classification Code
BAUA_Landmarks\tgr06087lpt.shp	Shapefile	TIGER Landmark Points	CFCC	Census Feature Classification Code
BAUA_Landmarks\tgr06095lpt.shp	Shapefile	TIGER Landmark Points	CFCC	Census Feature Classification Code
BAUA_Landmarks\tgr06097lpt.shp	Shapefile	TIGER Landmark Points	CFCC	Census Feature Classification Code
BAUA_Landmarks\tgr06099lpt.shp	Shapefile	TIGER Landmark Points	CFCC	Census Feature Classification Code

Data dictionary continued.

Layer	Type	Layer Description	Attribute	Description
BAUA_Landmarks\tgr06001lpy.shp	Shapefile	TIGER Landmark Polygons	CFCC	Census Feature Classification Code
BAUA_Landmarks\tgr06013lpy.shp	Shapefile	TIGER Landmark Polygons	CFCC	Census Feature Classification Code
BAUA_Landmarks\tgr06041lpy.shp	Shapefile	TIGER Landmark Polygons	CFCC	Census Feature Classification Code
BAUA_Landmarks\tgr06055lpy.shp	Shapefile	TIGER Landmark Polygons	CFCC	Census Feature Classification Code
BAUA_Landmarks\tgr06075lpy.shp	Shapefile	TIGER Landmark Polygons	CFCC	Census Feature Classification Code
BAUA_Landmarks\tgr06081lpy.shp	Shapefile	TIGER Landmark Polygons	CFCC	Census Feature Classification Code
BAUA_Landmarks\tgr06087lpy.shp	Shapefile	TIGER Landmark Polygons	CFCC	Census Feature Classification Code
BAUA_Landmarks\tgr06095lpy.shp	Shapefile	TIGER Landmark Polygons	CFCC	Census Feature Classification Code
BAUA_Landmarks\tgr06097lpy.shp	Shapefile	TIGER Landmark Polygons	CFCC	Census Feature Classification Code
BAUA_Landmarks\tgr06099lpy.shp	Shapefile	TIGER Landmark Polygons	CFCC	Census Feature Classification Code
BAUA_NHD1805\NHDH1805.mdb	Personal Geodatabase	National Hydrography Database		
NHDArea	Feature Class	Water Area polygons		
NHDLine	Feature Class	Rivers/streams polylines		
NHDWaterbody	Feature Class	Waterbody polygons		
NHDFlowline	Feature Class	Rivers/streams polylines direction of flow		
NHDPoint	Feature Class	Waterbody points		
BAUA_Orthoimagery\21319435	Raster Dataset	Digital Orthoimagery		

Exercise 3.1
Delineate DHS Urban Area

The first step in this chapter is to delineate the boundary of a DHS designated Tier 1 Urban Area. Then you will build a homeland security MEDS database as stipulated by the Department of Homeland Security.

Urban Areas are identified in the *FY 2007 Homeland Security Grant Program (HSGP)* document within the context of the Urban Areas Security Initiative (UASI), one of five homeland security grant programs.

> "The FY 2007 UASI program provides financial assistance to address the unique multidiscipline planning, operations, equipment, training, and exercise needs of high-threat, high-density Urban Areas, and assist them in building and sustaining capabilities to prevent, protect against, respond to, and recover from acts of terrorism."[13]

These Urban Areas are geographically defined as Tier 1 and Tier II areas. The HSGP applies a risk-methodology to determine the level of risk associated with high-density urban areas to include consideration of consequence, vulnerability, and threat. Intelligence assessments, population size and density, economic impacts, and proximity to nationally critical infrastructure, such as international borders, are all factors considered in this analysis.[14] Six Tier I urban areas, including the cities and jurisdictions around San Francisco; Chicago; Houston; Los Angeles/Long Beach; Washington, D.C.; and New York/New Jersey are eligible to share $420 million, or 55 percent of the total $747 million allocated to this program. The remaining 39 Tier II urban areas are eligible to share $336 million, or 45 percent of the allocated funds. This approach to determining geographic extent of high-risk areas beyond single boundaries emphasizes the regional approach homeland security has taken to securing areas beyond city boundaries.

The primary data layer of the MEDS is the boundary layer. This layer sets the spatial extent of the area under study. The MEDS framework identifies both Urban Area and Large Area spatial extents. Large Area extents refer to states or groups of states. Urban Area extents are spatially delineated by not just the civic boundary that marks the legal extent of a jurisdiction, but by combined entities encompassing additional areas that are part of the 10-mile buffer area beyond civic boundaries[15]. These combined entities make up the Urban Area Working Group (UAWG) and are eligible to apply to the DHS for funding within the context of an Urban Area Homeland Security Strategy drafted by the participating jurisdictions.

Tier 1 Urban Areas are the largest, most populated metropolitan centers in the country. The actual boundaries of these Urban Areas are set by mapping the extent of the combined entities in these metropolitan centers (e.g. contiguous counties, cities, townships, etc.), and the 10-mile buffer beyond.

In this tutorial, you will map the spatial extent of the San Francisco Bay Area Urban Area. The entities that make up this Urban Area are the 12 Bay Area cities with more than 100,000 people, which are Berkeley, Daly City, Fremont, Hayward, Oakland, Palo Alto, Richmond, San Francisco, San Jose, Santa Clara, Sunnyvale, and Vallejo, and a 10-mile buffer extending from the border of the combined area.[16]

The challenge for the cities that make up this Bay Area Urban Area (BAUA) is to coordinate collaborative efforts to promote interoperability in homeland security operations. This includes building integrated and comprehensive MEDS to align local capabilities with the National Incident Management Systems (NIMS), National Response Plan (NRP), National Infrastructure Protection Plan (NIPP), National Strategy for Transportation Security (NSTS), National Strategy for Maritime Security (NSMS), and the Implementation Plan for the National Strategy for Pandemic Influenza.[17]

This regional approach to homeland security operations realigns preparedness with more realistic boundaries, reinforcing the notion that terrorist threats and natural disasters don't recognize political boundaries.

However, the development of geospatial datasets has historically been limited geographically within political boundaries, as funding for data development has typically been granted to individual cities or counties. This has made for varying degrees of data consistency and integration between neighboring cities and jurisdictions. Regionalizing the pool of geospatial data to promote homeland security operations is an effective approach for enabling the nation's prevention, protection, response, and recovery regarding catastrophic events.

Open a new empty map document

1   Open a new empty map document.

2   Save the map document as **FL_GISHS_C3E1**, with FL being your initials, to **\ESRIPress\GISTHS \MYGISTHS_Work.**

Add boundary layer

1   From the Standard Toolbar, click the Add Data button.

2   In the Add Data dialog box, browse to **\ESRIPress\GISTHS\GISTHS_C3**.

3   Click the USA_Places.shp layer.

EXERCISES

4 Click Add.

The USA_Places.shp polygon layer is added to the map display. This layer is derived from the U.S. Census Bureau showing populated places that include census designated places, consolidated cities, and incorporated places within the United States.

View metadata

Each data layer is accompanied by a metadata file that defines key properties of the data. Metadata is often described as data about data. This information is stored in a file with the extension .xml and can be viewed within ArcMap, or edited and viewed in ArcCatalog. To view the metadata for the USA_Places.shp file,

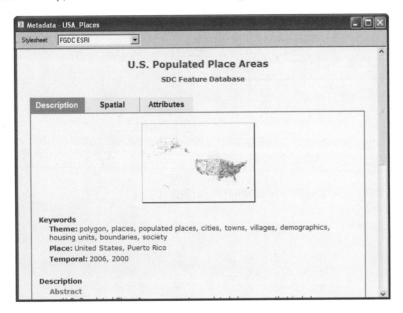

1 Right-click USA_Places in the Table of Contents, and select Data, then View Metadata.

The metadata style sheet contains a Description tab, a Spatial tab, and an Attribute tab. The Descriptive tab contains descriptive information about the dataset. The Spatial tab contains information about the spatial properties of the dataset such as its projection and coordinate extent. The Attribute tab contains a list of all of the attribute fields in the dataset and their properties.

EXERCISES

2   Scroll down the metadata window to view the descriptive information about this layer.

3   Click on any of the green links to toggle the information on or off.

4   Click the other tabs to review the spatial and attribute metadata.

5   Close the metadata window.

Select by Attributes

This USA_Places layer contains all of the census designated places, consolidated cities, and incorporated places within the United States. From this layer you can select out the places specific to your urban area.

1   From the Main Menu, click Selection, and Select by Attributes.

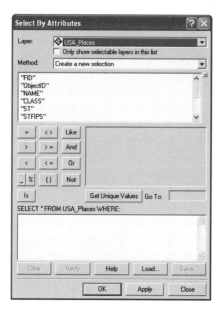

The Select By Attributes dialog box opens.

2   Double-click the "ST" field so that it appears in the selection box.

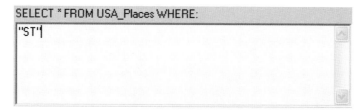

3   Click the Equals Operator ⟦ = ⟧ button so it appears in the selection box.

4   Click the Get Unique Values button to list all of the ST values in the database.

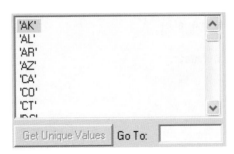

All of the state abbreviations appear in the list box.

5   Double-click 'CA' so that it appears in the selection box.

By first selecting only California places, you are ensuring that you do not include places in other states that have the same name as those in the designated DHS Urban Area in California.

6   Click Apply.

Only those places in the state of California are selected.

EXERCISES

7   Return to the Select By Attribute dialog box and select from the Method: drop-down list Select from current selection.

8   Click the Clear button to remove the "ST" selection expression.

9   Double-click the "NAME" field so that it appears in the selection box.

10 Click the Equals Operator button.

11 Click the Get Unique Values button to list all of the NAME values in the database.

12 Scroll down to 'Berkeley' and double-click to add it to the selection box.

Be sure to select 'Berkeley,' not 'Berkley,' as there is another place in the list with the same name, but spelled differently!

13 Click the OR button so that it appears in the selection box after 'Berkeley.'

14 Double-click the "NAME" field, and click the Equals Operator button again so that they appear in the selection box, and select 'Daly City.'

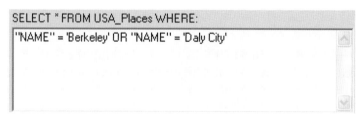

15 Repeat this selection process for the following places:
>   Fremont
>   Hayward
>   Oakland
>   Palo Alto
>   Richmond
>   San Francisco
>   San Jose (Not San José)
>   Santa Clara
>   Sunnyvale
>   Vallejo

SELECT * FROM USA_Places WHERE:

```
"NAME" = 'Berkeley' OR "NAME" = 'Daly City' OR "NAME" =
'Fremont' OR "NAME" = 'Hayward' OR "NAME" = 'Oakland' OR
"NAME" = 'Palo Alto' OR "NAME" = 'Richmond' OR "NAME" = 'San
Francisco' OR "NAME" = 'San Jose' OR "NAME" = 'Santa Clara' OR
"NAME" = 'Sunnyvale' OR "NAME" = 'Vallejo'
```

16 Click OK.

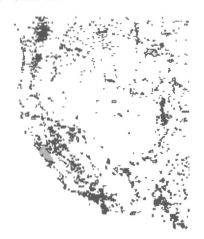

Only those places in California that are designated by DHS as places within the San
Francisco Urban Area are selected from the database and highlighted in bright blue.

17 From the Main Menu, click Selection, and Zoom to Selected Features.

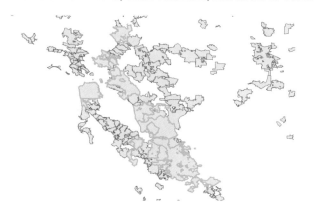

The map display zooms to the extent of only those selected features.

Create a new layer of a selected subset of features

Since you are only now interested in using this selected subset of place features in your GIS, you can create a new layer of just these features.

1 Right-click USA_Places and select Data, and then Data, Export Data.

2 Export Selected Features, using the same coordinate system as this layer's source data.

3 Save the output shapefile as **FL_SF_Places.shp** to **\ESRIPress\GISTHS\MYGISTHS_Work** with FL being your initials.

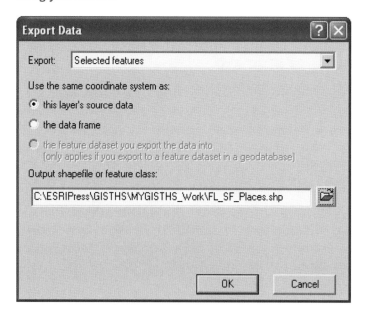

4 Click OK.

5 Click Yes to add the exported data to the map as a layer.

6 Uncheck USA_Places in the Table of Contents.

Only those places in California that are designated by DHS as places within the San Francisco Urban Area are now included in this new layer.

Project a data frame

You may notice when viewing this layer on the screen, that the spatial perspective of the data appears skewed, or not quite realistic. Some combination of area, shape, direction, and distance is distorted. This layer is mapped using the 3D geographic coordinate system of longitude and latitude. It is displayed as if it were peeled off of the spherical earth and stretched out onto a flat surface. In order for this layer to appear undistorted when viewed on a 2D surface, such as a computer screen or paper map, it needs to be projected to a coordinate system that accommodates its true 3D character.

Mathematical formulations are applied to layers to transform them from three dimensions to two. U.S. Census data layers are typically delivered in geographic coordinates. Data acquired from local agencies is usually in a projected coordinate system that accounts for the layer's 3D spatial properties at a large-scale extent.

ArcMap enables you to combine layers with different map projections into a map document. You can also convert your data to any number of projections and forms to view and analyze your data more accurately.

A common projected coordinate system is the universal transverse Mercator or UTM projection. This system divides the earth into zones to accommodate the curvature of the earth and minimize the local effects of distortion on area, shape, direction, and distance within each zone. California is in UTM zone 10. In addition, the MEDS specifications require that all data used for homeland security operations be in the North American Datum of 1983, which is a mathematical formulation that accounts for the true spherical shape of the earth.

1 Right-click the data frame Layers in the Table of Contents, and select Properties.

2 Click the Coordinate System tab.

EXERCISES

3   Expand the Predefined, Projected Coordinate System, UTM, NAD_1983 options.

4   Select NAD 83 UTM Zone 10.

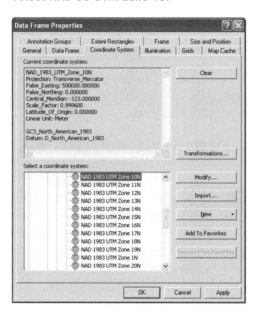

The Current coordinate system box is populated with the new parameters of this projected coordinate system.

5   Click OK. If a Warning box appears, click Yes to use this coordinate system.

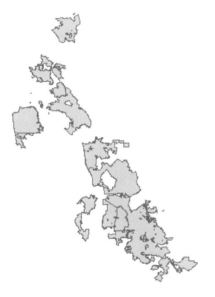

The map converts to this projected coordinate system and displays the layer in a more spatially accurate and realistic format.

Notice that the coordinates recorded at the lower right corner of the ArcMap window are now in meters. `585210.975  4148224.623 Meters` UTM uses meters as its unit of measurement.

EXERCISES

Project a data layer

Now that the data frame is converted to the UTM projected coordinate system, you can export this data layer and save it in this same coordinate system. While ArcMap is fully capable of managing layers in different coordinate systems, it is best to convert all of your data to a common system to eliminate any alignment or accuracy problems noted in the warning box that appeared when converting the data frame.

1   Right-click FL_SF_Places.shp in the Table of Contents, and select Data, and then Export Data.

2   Export All Features, using the same coordinate system as the data frame.

3   Save the output shapefile as **FL_SF_Places_UTM.shp** to **\ESRIPress\GISTHS\MYGISTHS_Work** with FL being your initials.

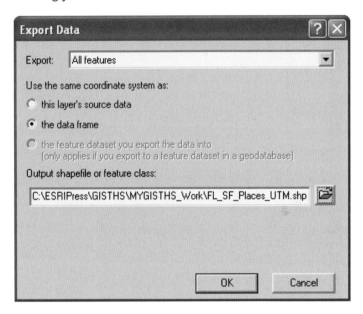

4   Click OK.

5   Click Yes to add the exported data to the map as a layer.

6   Right-click FL_SF_Places and click Remove button ✖ to delete this layer from the Table of Contents.

Dissolve feature boundaries

The DHS specifies that the Bay Area Urban Area includes a 10-mile buffer extending from the border of the combined area. The FL_SF_Places_UTM shapefile now contains 12 individual polygons, one for each feature, or city, in the designated urban area. To generate a 10-mile buffer beyond the border of the combined area, the individual city borders need to be dissolved to create one single feature, or polygon, of the combined extent of all 12 cities.

1 From the button bar, click the ArcToolbox button.

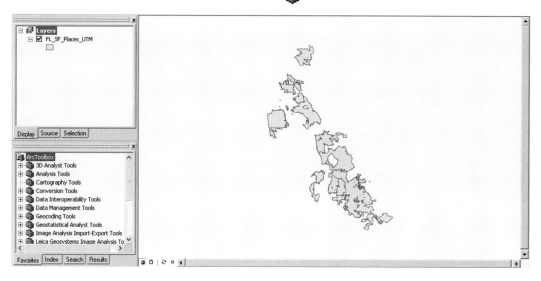

The ArcToolbox window opens and can be docked anywhere on the ArcMap screen.

2 Expand Data Management Tools and Generalization in the ArcToolbox window.

3 Double click Dissolve.

4 Select FL_SF_Places_UTM from the Input features drop-down list.

5 Accept the Output Feature Class to \ESRIPress\GISTHS\MYGISTHS_Work\FL_SF_Places_UTM_ Dissolve.shp.

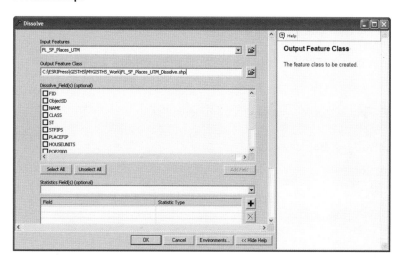

6   Click OK.

7   When the dissolve function is complete, click Close.

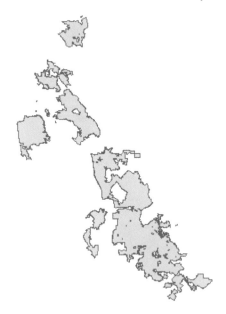

A new data layer is added to the map. This data layer contains only one feature, which is the combined spatial extent of all of the cities designated by DHS as the Bay Area Urban Area. There are no longer internal boundaries between the cities in the feature.

Generate buffer

Now that you have created a single feature of the combined spatial extent of all cities included in the DHS Bay Area Urban Area, you can generate a 10-mile buffer beyond this boundary to capture the full extent of the designated area.

1   Expand Analysis Tools and Proximity in the ArcToolbox window.

2   Double click Buffer.

3   Select FL_SF_Places_UTM_Dissolve from the Input features drop-down list.

4   Accept the Output Feature Class to **\ESRIPress\GISTHS\MYGISTHS_Work\FL_SF_Places_UTM_ Dissolve_Bu.shp**.

5   Type **10** as the Linear unit, and select Miles from the drop-down list.

6   Select ALL from the Dissolve Type drop-down list.

7   Click OK.

8   When the buffer function is complete, click Close.

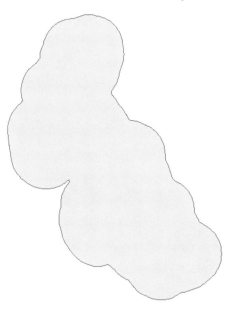

A new data layer is added to the map. This data layer contains only one feature, which is the spatial extent of a 10-mile buffer around all of the cities designated by DHS as the Bay Area Urban Area.

Adjust the transparency of a data layer

1   To view this new layer over the other layers, select this layer in the Effects toolbar.

2   Click the Adjust Transparency tool.

3 Drag the transparency bar up to 50%.

This buffer layer is now transparent enough to view its extent 10 miles beyond the combined borders of the DHS designated cities in the Bay Area Urban Area. This is the area for which DHS funding may be applied within the context of an Urban Area Homeland Security Strategy.

YOUR TURN

Refer to appendix D documents "DHS List of Fiscal Year 2006 Urban Areas Security Initiative (UASI) eligible applicants" and "USGS The Universal Transverse Mercator (UTM) Grid" to find a designated DHS area closest to your community.

Perform the select, project, and dissolve functions to the USA_Places shapefile to delineate the spatial extent of this Urban Area. Generate the 10-mile buffer beyond the combined border of the designated jurisdictions.

4 Save the map document when complete.

Exercise 3.2
Add boundary data to the MEDS map

Now that you have delineated the spatial extent of the DHS Bay Area Urban Area (BAUA), you can begin to add other layers of data specific to the MEDS. The MEDS data is grouped into categories specific to geographic sphere and include orthoimagery, elevation, hydrography, transportation, boundaries, structures, land cover, and geographic names. These data layers provide a basemap to build and manage an effective homeland security operation.

Rename data frame

Similar to layers, you can rename a data frame to a custom title that gives it a more specific reference to your work.

1 Open a new ArcMap document.

2 Save this map document as **FL_GISHS_C3E2**, with FL being your initials, to **\ESRIPress\GISTHS \MYGISTHS_Work**.

3 Right-click the data frame Layers in the Table of Contents, and select Properties.

4 Click the General tab.

5 Type **BAUA MEDS** in the Name field.   Name: BAUA MEDS

6 Click OK.

The data frame is now named BAUA MEDS.   BAUA MEDS

Create a group layer

You can also group several layers into a "group layer" that can be turned on and off with one click of the mouse. It also helps to organize the processing and presentation of the data. In the case of the MEDS data, it is helpful to group each category of data according to its MEDS classification. The first set of layers to add to the map shows the boundaries that delineate the BAUA and surrounding extent.

1 Right-click the data frame BAUA MEDS, and select New Group Layer.

2 Right-click New Group Layer, and select Properties.

EXERCISES

3 Type **Boundaries** in the Layer Name field.

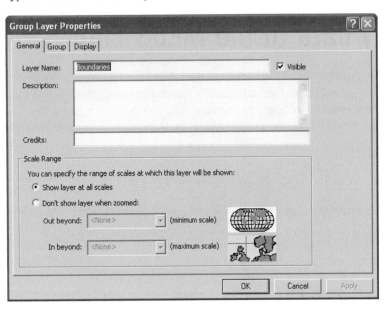

Add layer in Group Layer Properties box

Existing layers can be added to a new group layer by listing them in the Group Layer Properties box.

1 Click the Group tab.

2 Click the Add button, and select FL_SF_Places_UTM_Dissolve_Bu.shp from the **\ESRIPress\ GISTHS\MYGISTHS_Work** folder.

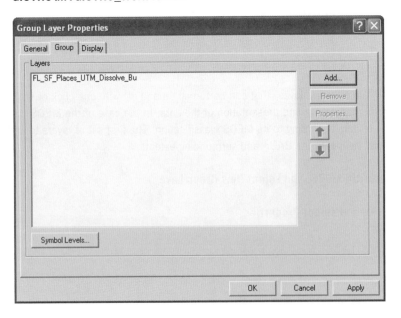

3 Click OK.

This layer is added to the Boundaries group layer, and is nested below the group layer name in the table of contents.

4 Rename FL_SF_Places_UTM_Dissolve_Bu to **BAUA Boundary**.

Select by location (intersect)

To add spatial context to this layer, you will add county and state layers to the map document. Since you don't need the boundaries of all of the counties and state in the country, you will select only those features that intersect with this BAUA boundary.

1 Add USA_Counties.shp from **\ESRIPress\GISTHS\GISTHS_C3**.

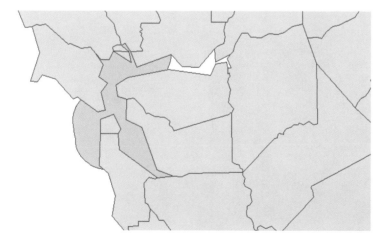

This layer is added outside of the BAUA MEDS group layer, and overlays the BAUA boundary layer.

EXERCISES

2   From the Main menu, select Selection, and click Select by Location.

3   Check the box next to USA_Counties to select features from.

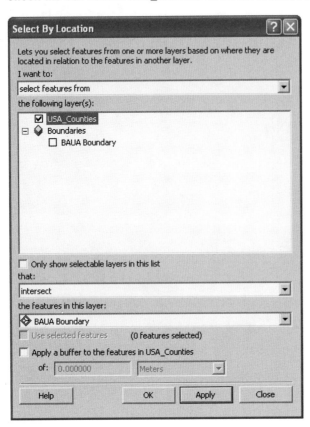

4   Select intersect and BAUA Boundary.

5   Click OK.

Only those counties that intersect with the BAUA boundary are selected.

6   Open the USA_Counties attribute table and view only the selected counties.

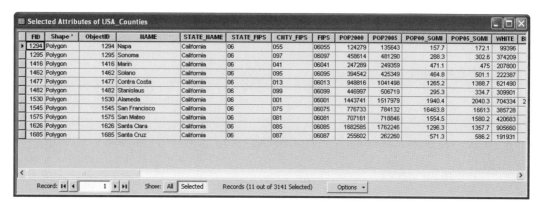

There are 11 out of 3,141 counties in the United States that are part of this BAUA.

7   Close the attribute table.

8   Export these selected features to a new shapefile using the same coordinate system as the data frame, and name it **FL_BAUA_Counties.shp**, with FL being your initials, in the **\ESRIPress\GISTHS \MYGISTHS_Work folder.**

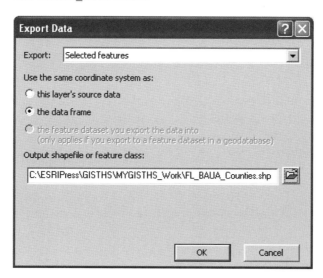

9   Click Yes to add it to the map as a layer.

10   Drag **FL_BAUA_Counties** into the Boundaries group layer above BAUA Boundary.

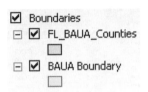

11   Remove **USA_Counties** from the Table of Contents.

12   Zoom to the extent of the **FL_BAUA_Counties** layer.

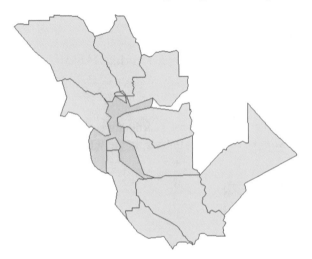

Change symbology of polygon feature

To examine the extent of the county boundaries over the BAUA boundary layer, you need to change the polygon fill symbology to enable viewing.

1   Click the polygon symbol under the **FL_BAUA_Counties** layer name in the Table of Contents.

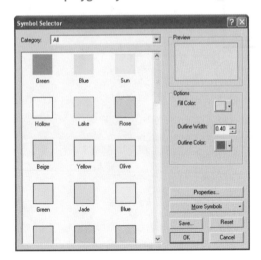

The Symbol Selector window opens.

2   Click Hollow as the fill.

3   Increase the Outline Width to 1.5.

4   Outline Color Select Gray 50%.

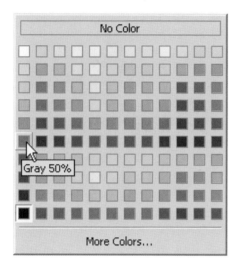 

5   Click OK.

The extent of the BAUA is now visible beneath the BAUA_Counties layer.

6   Label the BAUA_Counties data layer using NAME as the label field.

Add layer to group layer

You can also add layers to a group layer from the table of contents.

1   Right-click the Boundaries group layer title.

2   Click Add Data. ✛

3   Select FL_SF_Places_UTM.shp from **ESRIPress\GISTHS\MYGISTHS_Work**.

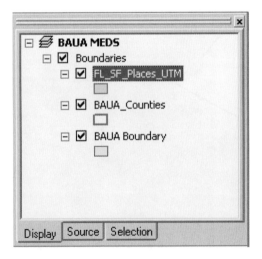

This layer is added directly into the group layer and displayed on the map.

Note that in this case, this spatial analysis is critical to identifying the county partners designated to collaborate as part of a homeland security Urban Area Working Group (UAWG). There are no BAUA cities in Marin, Sonoma, Napa, Santa Cruz, or Stanislaus counties, but because the 10-mile buffer beyond the combined border of the BAUA cities extends into these counties, they are included in the area covered by DHS programs.

These combined city and county entities make up the Bay Area Urban Area Working Group and are eligible to apply to the DHS for funding within the context of an Urban Area Homeland Security Strategy drafted by the participating jurisdictions. The Bay Area Regional GIS Council was formed in 2002 to support regional GIS coordination and identify and encourage data sharing among jurisdictions within the BAUA.[18] Its proposed Homeland Security Data Server/Project Homeland initiative would build a system of data servers to enable sharing of critical infrastructure, key asset data, and information across governmental jurisdictions to help organizations publish, discover, access, exchange, and maintain the vital geospatial information required to support critical infrastructure protection, response, and recovery operations.[19]

Organizations around the country similar to the Bay Area Regional GIS Council are working toward creating redundant node-based data repositories for storing and distributing local and regional geospatial data. These centers will be instrumental in providing critical data and information to homeland security planners and first responders. Many of these centers are still in the planning stages and are expected to provide the model for future data storage, analysis, and distribution needs.

YOUR TURN

> Add the USA_States.shp layer from the \ESRIPress\GISTHS\GISTHS_C3 folder, and use either the Select By Location or Definition Query to create a new layer of only the boundary of California. Add it to the Boundaries group layer.

4   Save the map document when complete.

Exercise 3.3
Add transportation data to the MEDS map

Change the data frame display units

The geographic extent of the BAUA is not a designated political boundary but rather a combined boundary extended by a 10-mile buffer. In order to define this area to download data from the National Map Seamless Server (NMSS), a bounding box delimited by four x,y longitude and latitude coordinates needs to be created. Since your data frame is projected into UTM coordinates, you will change the map display units to decimal degrees to record these coordinates.

1   Right-click the BAUA MEDS data frame in the Table of Contents, and select Properties.

2   Click the General tab and select Decimal Degrees from the Units Display: drop-down list.

3   Click OK.

The map display coordinates in the lower right corner of the ArcMap window are now shown in decimal degrees.

4   Save this map document as **FL_GISHS_C3E3**, with FL being your initials, to **\ESRIPress\GISTHS\ MYGISTHS_Work**.

Define spatial extent of BAUA

1   Zoom to the extent of the BAUA Boundary layer.

2   Set your cursor at the upper extent of the BAUA Boundary layer.

3   Note the longitude and latitude coordinates for this location shown at the bottom right of ArcMap window. `-122.212  38.321 Decimal Degrees`

Note: Your coordinate values may be slightly different.

4   Record only the latitude coordinate in a notepad file or on a scrap of paper, noting the location of this position as Top.

5   Repeat this process for the remaining three extents of the BAUA Boundary layer, recording the farthest latitude of the bottom position, and farthest longitude of the right and left positions

Note: Your coordinate values may be slightly different.

Top	38.321
Right	-121.412
Bottom	36.979
Left	-122.696

Download transportation data from the National Map Seamless Server

Now that you have created a basemap of jurisdictional boundaries of the BAUA, you can begin to populate your GIS with additional MEDS data from the National Map via the National Map Seamless Server.

1   Open your Web browser and navigate to **http://seamless.usgs.gov/**.

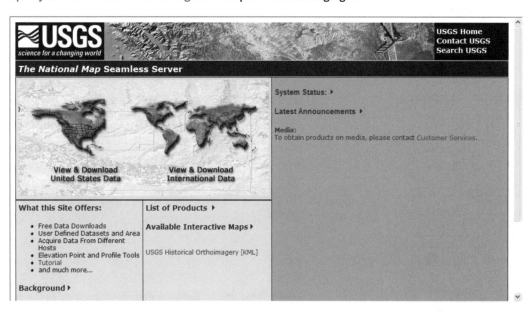

2   Click the link View and Download United States Data.

The USGS Seamless Data View window opens. It may take a few minutes to load all of the viewer's tools and features.

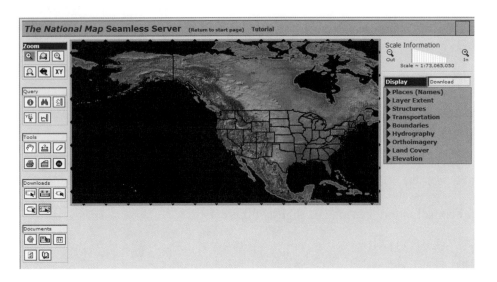

The viewer opens with the GTOPO60 Global Color Shaded Relief layer displayed.

3   Familiarize yourself with the tools and features of the viewer by scanning your cursor over each tool to show its description.

Note: Screen features and buttons may appear slightly different depending on the browser you are using.

The **Zoom** tools enable you to zoom in to the full extent, out to a region or an area, to the previous extent, or to an XY point on the map.

The **Query** tools enable you to display the map feature attributes, search for a geographical place, or query using a set of elevation or U.S. Grid tools.

The **Tools** enable you to pan the map display, measure features, print to PDF, view metadata, or set layer transparency.

The **Downloads** tools enable you to define a download area by dragging the mouse on the map display, by using single or multiple templates, or by setting the spatial extent by entering bounding coordinates.

The **Documents** tools enable you to view map and metadata information, set the map legend, refresh the map display, and access the user instructions.

4 View the Scale and Layers setting to the right of the viewer.

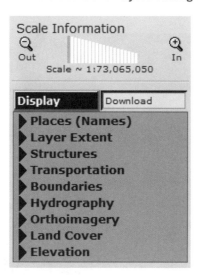

The scale is set to the farthest zoomed out range to show the entire United States geographic extent.

The layers under the Display tab show the list of group layers available for viewing in the map display. Note that this list closely aligns with the DHS MEDS list.

5 Click the Download tab.

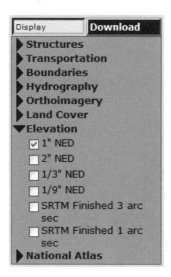

The layers under the Download tab show the list of group layers available for download to the user. Note that this list, though not exactly the same as the Display list, also closely aligns with the DHS MEDS list.

6   Click the Zoom In button.

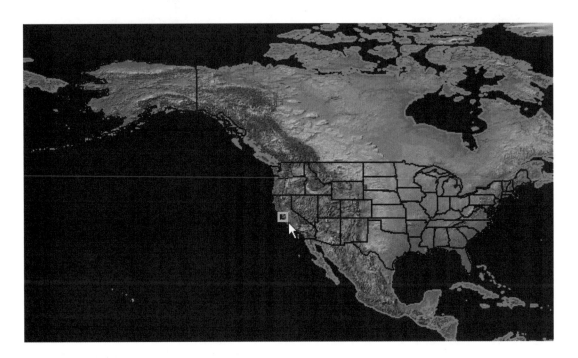

7   Drag the cursor to draw a box around the BAUA extent, and click.

The map display zooms to the spatial extent of the BAUA, and shows a set of elevation,

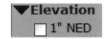

transportation, and boundary layers set to display at this scale.

8   Click to uncheck the 1" NED dataset under the Elevation group layer.

9   Expand the Transportation group layer and click to put a check in the box next to BTSRoads.

10   Click the Define Download Area by Coordinates tool in the Downloads Tools group box.

11   Click at the bottom of that window to Switch to Decimal Degrees.

12   Type in the longitude of the left and right coordinates, and the latitude of the top and bottom coordinates recorded earlier to define the spatial extent of the BAUA.

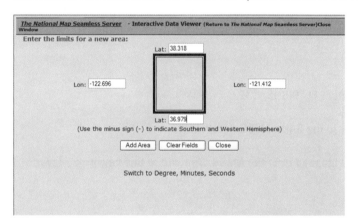

13   Click Add Area.

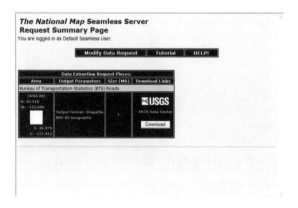

The National Map Seamless Server Request Summary Page appears listing the requested dataset.

14 Click the Download button.

The National Map Seamless Server

Current order status

Adding your request to the queue...

Please wait for the data to be returned.

The Current order status window appears as the data is extracted.

15 When the order is ready, save the zip file to **\ESRIPress\GISTHS\MYGISTHS_Work** and extract it using a zip file program. **Please be patient, as this download process may take a few minutes**

Note: If you have limited Internet connectivity, this file is available locally at \ESRIPress \GISTHS\GISTHS_C3\Downloaded_Data_C3\BAUA_BTSRoads. Copy it into your \ESRIPress\GISTHS\MYGISTHS_Work folder before you begin working with it.

Add a Transportation group layer

1 Return to the ArcMap map document **FL_GISHS_C3E3**.

2 Add a Transportation group layer to the **BAUA MEDS** data frame.

3 Add the BTS Roads shapefile downloaded from the NMSS Web site to this new group layer.

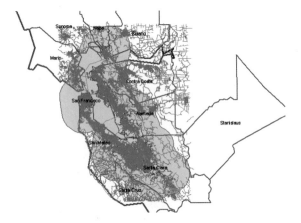

The road shapefile is drawn to the map display to the extent of the coordinates used to define the spatial area of the BAUA.

Clip layer to boundary extent

The BAUA spatial extent is defined as a bounding box of four coordinates, though it is truly an irregular shape delineated by the city boundaries and buffers. The transportation data downloaded from the NMSS extends to the coverage of this bounding box, including data beyond the actual boundary of the BAUA. This is a very complex dataset that contains many records of line segments. Any attempt to reduce the amount of data included in this dataset will benefit your GIS. Clipping the transportation data layer to the actual extent of the BAUA boundary will enhance the performance of your GIS as it processes and analyzes geospatial information.

1   Open the ArcToolbox window if it is not already open.

2   Expand Analysis Tools and Extract in the ArcToolbox window.

3   Double click Clip.

4   Select your roads data layer from the Input features drop-down list.

5   Select the BAUA Boundary data layer in the Boundaries group layer as the feature to use to clip the input feature.

6   Output Feature Class to **\ESRIPress\GISTHS\MYGISTHS_Work\FL_BAUA_Roads_Clip.shp**, with FL being your initials.

7   Click OK.

8   Close the processing window when the clip is complete.

9   Remove from the Table of Contents the transportation data layer that you downloaded from the NMSS.

10 Export the FL_BAUA_Roads_Clip data layer to **\ESRIPress\GISTHS\MYGISTHS_Work\ FL_BAUA_ Roads_UTM** using the same coordinate system as the data frame.

11 Remove FL_BAUA_Roads_Clip from the Table of Contents.

12 Drag the new FL_BAUA_Roads_UTM data layer into the Transportation group layer.

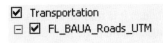

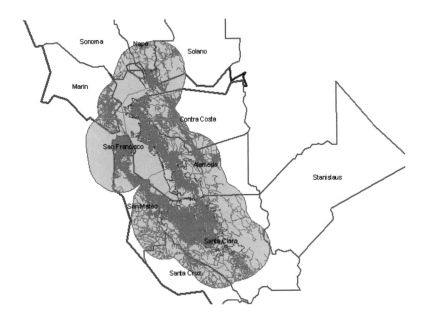

The new FL_BAUA_Roads_UTM data layer is added to the map display. It extends only to the boundary of the BAUA and is projected in the same coordinate system as the other data layers in the data frame. Preparing your data in a common projection and datum enables quick, efficient, and accurate data processing when needed. This is especially important in the case of a major disaster or catastrophic event when time is critical and the safety and security of people and places are in jeopardy.

Join tables to categorize feature classes

All road features are categorized on the map in the same way using a single line symbol. We know that not all roads are the same: some are local two-lane neighborhood roads that have little homeland security significance, while others are multilane interstate highways that are the economic lifeline and circulation routes for metropolitan areas around the country.

Each road is classified by its features and given a Feature Class Code (FCC) ranging from A00 to A75, to include primary, secondary, and minor crossings. Mapping the roads in the BAUA by FCC code helps homeland security planners identify roads that have the greatest impact on a region's security and welfare if compromised, as well as their value to evacuation planning.

1 Open the FL_BAUA_Roads_UTM attribute table.

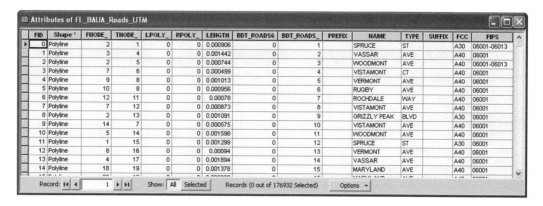

2 Scroll to the right and down the table to review the FCC codes.

The FCC codes do not describe the road. This information is in another table, cfcc.dbf. ArcMap can contain tabular data that is not spatial data, but provides additional attribute information about a feature.

3 Close the FL_BAUA_Roads_UTM attribute table.

4 Click the Add Data button.

5 In the Add Data dialog box, browse to **\ESRIPress\GISTHS\GISTHS_C3**.

6 Click cfcc.dbf.

7 Click Add.

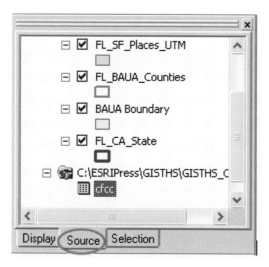

The table of contents automatically switches to the source view, showing the cfcc.dbf table, and all the data layers sorted by source or path.

8   Right-click cfcc.dbf to open the table.

9   Scroll to the far right to view the DESCRIPTIO and FCC fields.

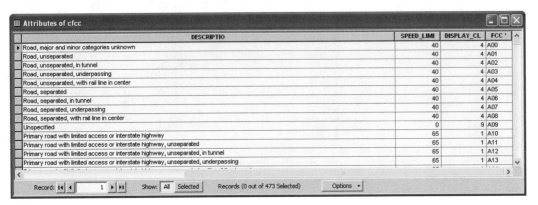

This table provides detailed road attribute data information to classify the BAUA roads by type.

10 Close the cfcc.dbf table.

11 Click the Display tab at the bottom of the Table of Contents.

12 Right-click FL_BAUA_Roads_UTM and select Joins and Relates.

13 Select Join to open the Join Data dialog box.

14 Select Join attributes from a table, and choose FCC as the field in this layer that the join will be based on.

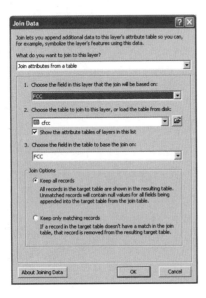

The other fields will automatically fill to select cfcc as the table to join and FCC as the field in that table to base the join on. The Join Option to Keep all records is also selected.

15 Click OK.

16 If asked, click Yes to automatically create an index for the join field.

17 Open the FL_BAUA_Roads_UTM attribute table and scroll to the far right.

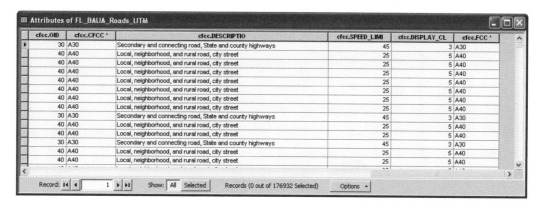

Each record is now appended with the cfcc table to include the description of the FCC code. Each road can now be categorized by this description to better identify its type and importance to homeland security operations and planning.

18 Close the FL_BAUA_Roads_UTM attribute table.

Categorize by unique value

1 Right-click FL_BAUA_Roads_UTM and select Properties.

2 Click the Symbology tab.

3 Click Categories / Unique Values, many fields from the Show window.

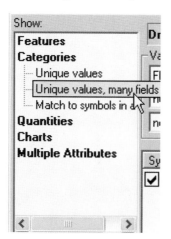

4   Select CFCC as the first field, and DESCRIPTIO as the second field from the Value Fields drop down list.

5   Click Add All Values button.

6   Click OK.

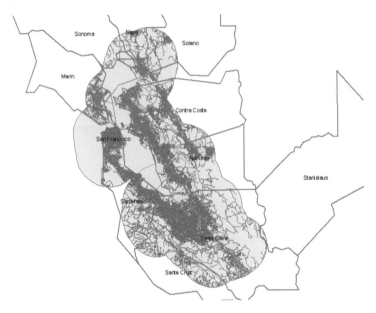

All of the roads are drawn using the same line width but with different colors for the line symbols. It is not possible to differentiate them using this symbol selection. By grouping and symbolizing them according to CFCC code and description, they are more easily delineated for visual and planning purposes.

7   In the Table of Contents, double-click the first road category symbol for A10 Primary road with limited access. Expand the column width if it is truncated.

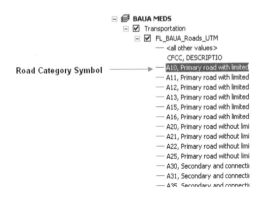

8 From the Symbol Selector, click More Symbols.

9 Click Transportation.

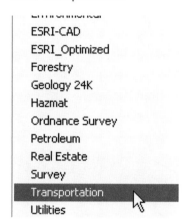

A new set of symbol markers used for transportation mapping is added to the Symbol Selector.

10 Select the CFCC Road Classification category.

11 Click the symbol for A10.

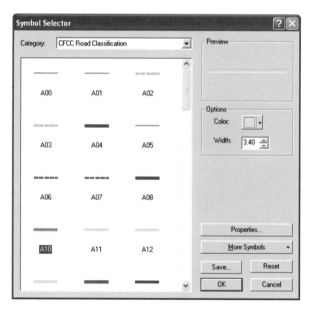

12 Click OK.

13 Repeat this process for A11, A12, and A13.

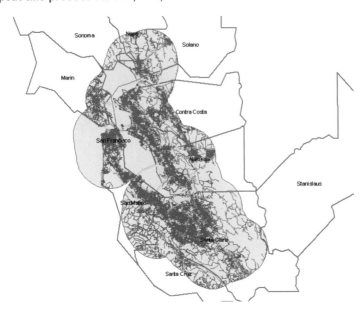

These primary roads are now drawn using a brighter and heavier line symbol and are more clearly depicted in the map display.

14 Zoom into a local area on the map to view the roads more clearly.

YOUR TURN

Categorize remaining roads using the CFCC Road Classification symbols in the Symbol Selector window. Use common symbols for those codes that do not have unique symbols: A62, A63, A70, A71, and A72.

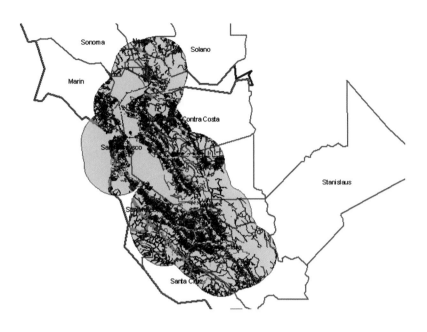

15 Zoom into a local area on the map to view the roads more clearly.

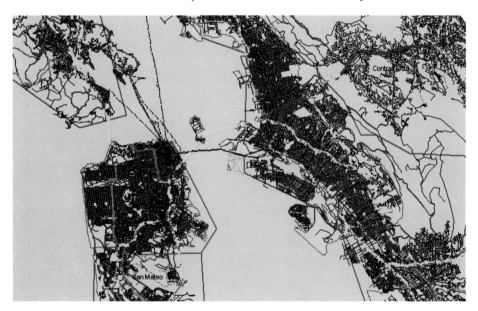

Now that you have been successful at downloading, clipping, categorizing, and symbolizing this transportation layer, it can be used in later chapters as a critical GIS data layer in homeland security planning and operations.

EXERCISES

Export roads by CFCC classifications

When a feature is complex and contains many points, lines, or small polygons, it may be difficult to display on a map at certain scales, making it impossible to read or use. ArcMap enables the display of features and labels at specified scales to show only when they are readable and relevant. The local roads in the current roads data layer is not effectively displayed when zoomed out to the full extent of the data layer. It is more useful to set this data layer and its labels to display to a more readable and useful larger scale extent. To do this, the roads data layer needs to be divided and exported by primary, secondary, and local classifications.

1   From the main menu, select Select by Attributes.

2   Select only the local roads from **FL_BAUA_Roads_UTM** containing the CFCC codes greater than or equal to A40 and less than or equal to A45. Be sure to use the **Create a new selection** as the selection method.

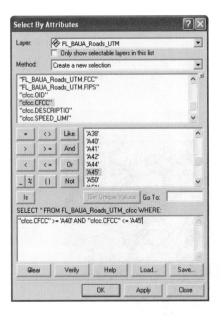

3   Click OK.

4   Zoom to the extent of the BAUA Boundary data layer.

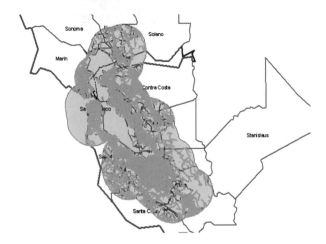

All of the local roads in the BAUA Roads data layer are highlighted. Creating a new layer of just these local roads enables you to set a visible scale to them so that they will only show when zoomed in to a readable and useable extent.

5 Right-click FL_BAUA_Roads_UTM, and then Data, Export Data.

6 Export Selected Features, using the same coordinate system as the data frame.

7 Output the shapefile to **\ESRIPress\GISTHS\MYGISTHS_Work\FL_BAUA_Roads_Local.shp**, with FL being your initials.

8 Click OK.

9 Click Yes to add the exported data to the map as a data layer.

10 From the Main menu, click Clear Selected Features.

11 Drag FL_BAUA_Roads_Local into the Transportation group layer, and turn off FL_BAUA_Roads_UTM.

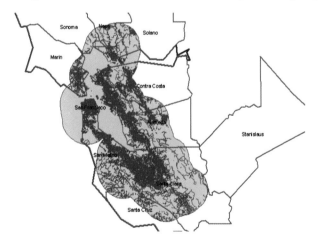

Only the local roads are displayed on the map.

Import symbology

ArcMap enables the user to import a symbology set assigned to another data layer to a new data layer that contains the same or subset of features. This eliminates the tedious task of repeatedly assigning the same symbology set to multiple data layers.

1 Right-click FL_BAUA_Roads_Local and select Properties.

2 Click the Symbology tab.

3 Click Categories / Unique Values, many fields from the Show window.

4 Select CFCC and DESCRIPTIO as the Value Fields.

5 Press Apply.

6 Click the Import button in the Layer Properties window.

7 In the Import Symbology dialog box, select FL_BAUA_Roads_UTM.

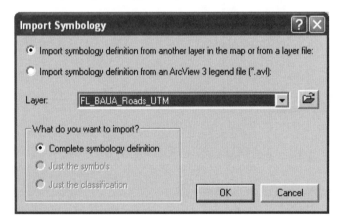

8 Click OK.

9 In the Import Symbology Matching Dialog box, select CFCC and DESCRIPTIO as the Value Fields, if they are not already preloaded.

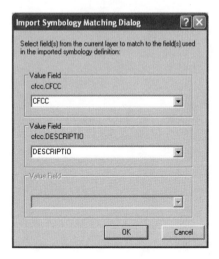

10 Click OK.

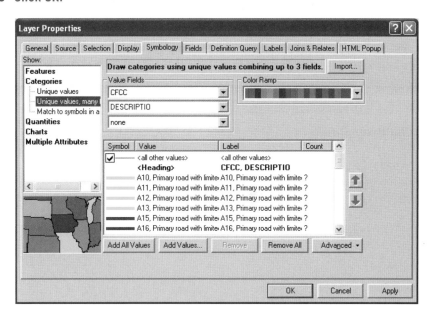

This imports the symbology set from the FL_BAUA_Roads_UTM data layer to this subset of data.

11 Click OK.

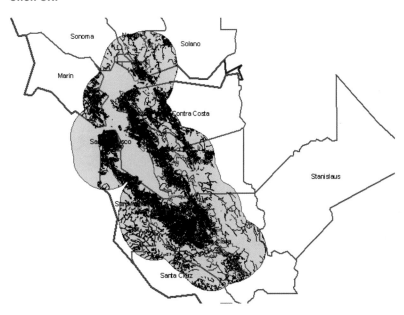

Only the local roads data layer is drawn to the map. It is using the symbology set imported from the FL_BAUA_Roads_UTM data layer.

YOUR TURN

Create a new data layer of only the primary roads (A10-A25); and another data layer of only the secondary roads (A30-A38). Apply the designated CFCC Road Classification symbols to each of the road categories by importing the symbology set from FL_BAUA_Roads_UTM to each of them.

Show features at specified scale range

1   Right-click FL_BAUA_Roads_Local and select Properties.

2   Click the General tab.

3   In the Scale Range box, click Don't show layer when zoomed: and select Out beyond 1:100,000 from the drop-down list.

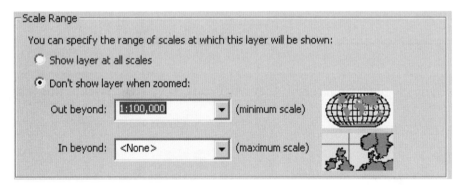

4   Click OK.

5   From the Main menu, select 1:100,000 from the scale box.

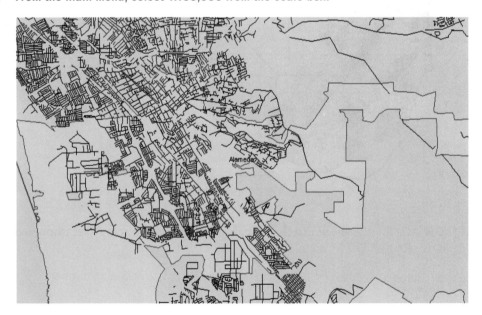

This data layer is visible in the map display.

6   Change the scale to 1:100,001.

This data layer is no longer visible.

YOUR TURN

> Set the minimum scale of the secondary roads data layer to 1:250,000 layer. Leave the primary roads layer to show at all scales.

Show labels at specified scale range

Adding labels to the roads data layer is helpful for identifying location. However, labels will overlap at too small a scale over a large area, making the map difficult to read. By scaling the labels, they are set to show only at a maximum scale so that they are most legible and helpful.

1   Right-click FL_BAUA_Roads_Local and select Properties.

2   Click the Labels tab.

3   In the Other Options box, click Scale Range.

4   In the Scale Range box, click Don't show labels when zoomed: and select Out beyond 1:24,000 from the drop-down list.

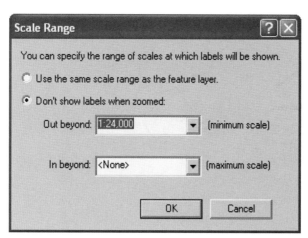

5   Click OK.

6   Click to check Label features in this layer.

7   Click OK.

8   From the Main menu, select 1:24,000 from the scale box.

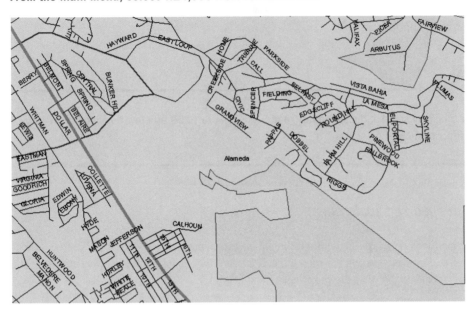

The road name labels are visible in the map display.

9   Change the scale to 1:24,001.

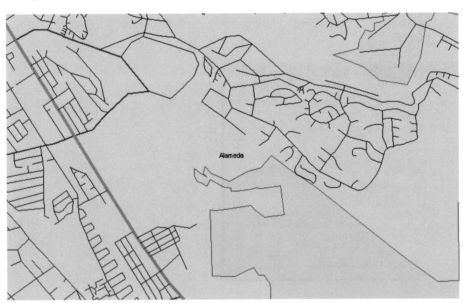

The road name labels are no longer visible.

Change label text symbol

All of the labels for the roads are defaulted to display at an Arial 8 font setting. When all of the roads are displayed on the map together, this makes it difficult to distinguish which labels are for which roads. ArcMap enables you to set different labels text symbols for different data layers to make the map more readable.

1   From the Main menu, select 1:24,000 from the scale box.

2   Right-click FL_BAUA_Roads_Secondary and select Properties.

3   Click the Labels tab.

4   Choose an alternate color and size for this label.

5   Click to check Label features in this layer.

6   Click OK.

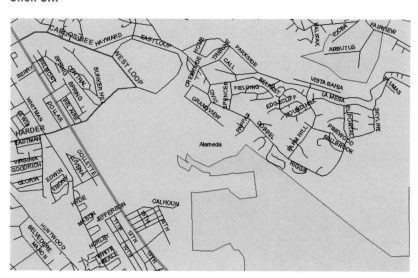

The labels for the secondary roads are now drawn to the map in a different font, and are more easily distinguished from the local road labels.

YOUR TURN

Change the text label symbol for the BAUA Counties data layer so it is more distinguishable when the map is zoomed to a scale of 1:100,000.

EXERCISES

Add more transportation layers

In addition to roads, other transportation systems include railroads, airports, and ferry crossings. Each of these transportation features plays a critical role in the day-to-day lives of many Americans, and because of their importance, are attractive targets to terrorists. Railroads are used to transport passengers to and from their workplaces and consumer commodities to distribution centers. They are also used to transport highly toxic and volatile substances to and from industrial centers and refineries. These trains pose a significant public safety and security threat to communities as they pass almost unnoticed through the backyards of the country.

Some of these data layers have already been downloaded for use in this chapter. You will create another as a subset of the BTS dataset downloaded from the NMSS. You will add them all to the BAUA_MEDS map to complete the minimal essential transportation dataset.

1 From **\ESRIPress\GISTHS\GISTHS_C3**, add BAUA_Railoads.shp to the Transportation group layer.

2 Change the symbology to the Railroad line symbol.

3 Add BAUA_Airports.shp to the Transportation group layer.

4 Change the symbology to the Airport marker symbol.

5 Select and export a subset of FL_BAUA_Roads_UTM.shp containing only A65 ferry crossings, and save it to **\ESRIPress\GISTHS\MYGISTHS_Work\FL_BAUA_FerryCrossing.shp**, with FL being your initials.

6 Add the new Ferry Crossing data layer to the Transportation group layer.

7 Change the symbology of the Ferry Crossing to the CFCC Code Classification A65 line symbol at line width 4.00.

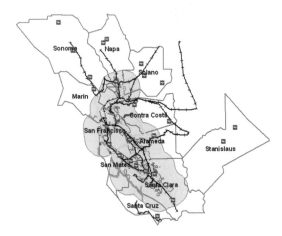

The Transportation group layer now contains a set of data layers that define the movement of people, goods, and services in this urban area. Surveillance and analysis of this data is critical to successful homeland security planning and operation.

8 Save the map document when complete.

Exercise 3.4
Add hydrography data to the MEDS map

We depend on our nation's waterways in many critical ways. They provide vital transportation corridors for passengers, goods, and services. Our rivers, streams, reservoirs, and aqueducts sustain life by storing and preparing our water for drinking, irrigation, and manufacturing commodities we use every day. Much of our public safety depends on the protection of this national resource.

Download National Hydrography Database (NHD) geodatabase

The USGS and U.S. Environmental Protection Agency (EPA) have collaborated to compile the National Hydrography Database (NHD) as a component of the National Map. The NHD is a comprehensive set of digital spatial data that contains information about surface water features such as lakes, ponds, streams, rivers, springs, and wells.[20] It is available at high, medium, and low resolution, and contains an extensive set of point, line, and polygon features characterizing the full extent of water resources and facilities across the nation.

Up to now, you have worked with geospatial vector data in the shapefile and tabular formats. These are not the only data structures ArcMap is capable of handling. Other formats include the raster data structure, explored later in the chapter, used to display aerial and satellite images, as well as surface models of continuous data representing such features as elevation and climate data. The NHD is delivered in yet another ArcMap data structure known as the *geodatabase*. The geodatabase is defined as "...a collection of geographic datasets of various types held in a common file system folder, a Microsoft Access database, or a multiuser relational database (such as Oracle, Microsoft SQL Server, or IBM DB2)."[21]

The geodatabase can contain three primary dataset types:

- Tables
- Feature classes
- Raster datasets

All three primary datasets in the geodatabase, as well as other geodatabase elements, are stored using tables. Once the geodatabase is designed and populated with data, it can be enhanced with more advanced capabilities (such as by adding topologies, networks, or subtypes) to model GIS behavior, maintain data integrity, and work with an important set of spatial relationships.[22]

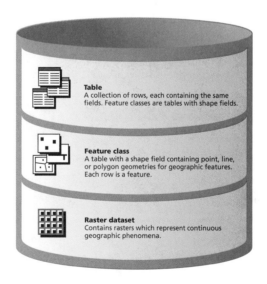

Table
A collection of rows, each containing the same fields. Feature classes are tables with shape fields.

Feature class
A table with a shape field containing point, line, or polygon geometries for geographic features. Each row is a feature.

Raster dataset
Contains rasters which represent continuous geographic phenomena.

The NHD geodatabase contains tables and feature classes that define the full extent of water resources in the United States. The feature classes contain point, line, area, flow line, and water body data for each basin, region, and watershed hydrologic unit.

1   Navigate your Internet browser to http://nhd.usgs.gov/.

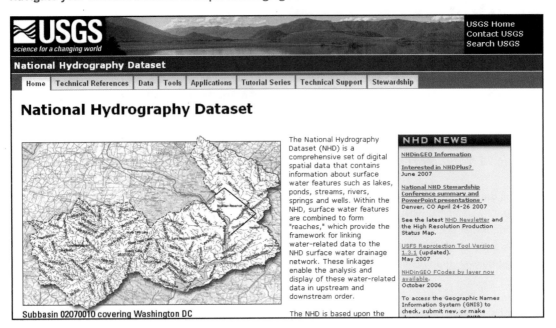

2   Click the Data tab, the top bar of the Web page. | Data |

3   Click Go to the NHD viewer.   Go to the NHD viewer

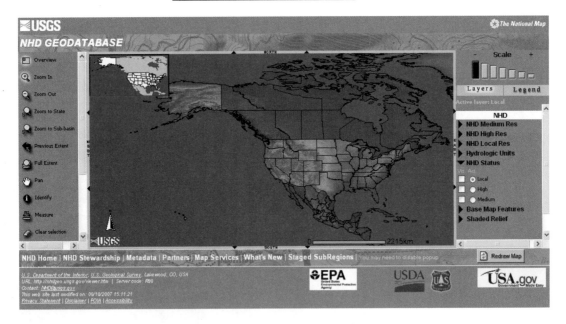

The NHD Geodatabase Viewer opens showing a map screen similar to the NMSS.

4   Click the Zoom In tool.   Zoom In

5   Drag a box around the approximate extent of the BAUA.

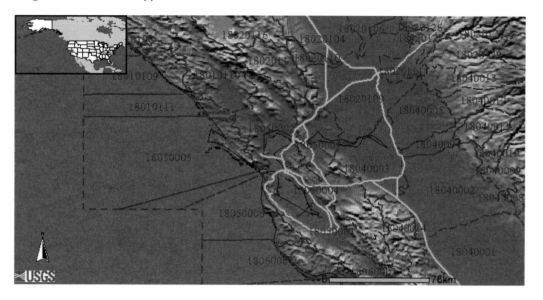

The viewer zooms to the approximate extent of the BAUA showing the extent of each subregion. The BAUA is located in the 1805 subregion.

6   Click Staged Subregions.   **Staged SubRegions**

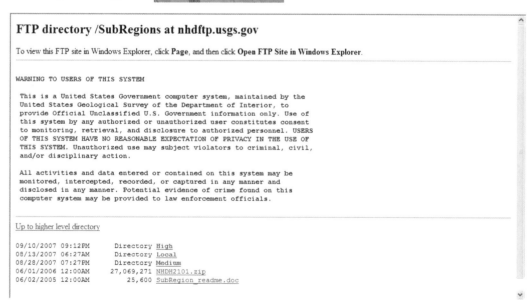

The NHD FTP server page appears.

7   Click High.

8   Scroll down and click NHD1805.zip.

9 Save and extract this zip file to **\ESRIPress\GISTHS\MYGISTHS_Work**.

Note: If you have limited Internet connectivity, this file is available locally at \ESRIPress \GISTHS\GISTHS_C3\Downloaded_Data_C3\BAUA_NHD. Copy it into your \ESRIPress\GISTHS\MYGISTHS_Work folder before you begin working with it.

10 Open ArcCatalog and navigate to **\ESRIPress\GISTHS\MYGISTHS_Work**.

11 Expand the **NHD1805.mdb** geodatabase, and the **Hydrography** and **HydrologicUnits** Feature Datasets.

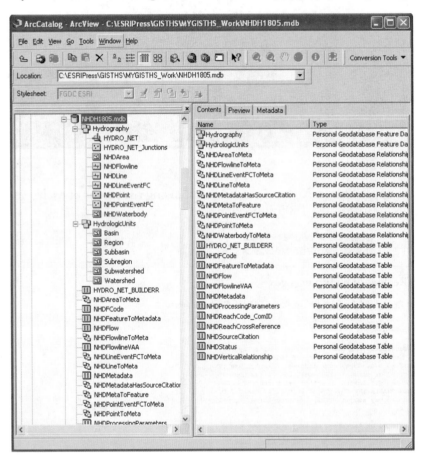

The Hydrography Feature Dataset is composed of the primary NHDArea, Flowline, Line, Point, and Waterbody feature classes. The _NET and EventFC files are linear referencing files associated with a feature class geometric network.

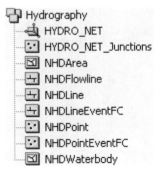

The Hydrologic Units are the polygon feature classes delineating the spatial extent of each unit at the basin, region, subbasin, subregion, subwatershed, and watershed levels.

The remaining files are tables and relationship classes that define the features and their associations to each other. [23]

Add features of the hydrography geodatabase to the MEDS map

1   Open **FL_GISHS_C3E3.mxd**, if it is not already open.

2   Turn off all layers except FL_BAUA_Counties.

3   Zoom to FL_BAUA_Counties layer.

4   Create a Hydrography group layer in the BAUA MEDS data frame.

5   From ArcCatalog, drag the NHDArea feature class into the Hydrography group layer.

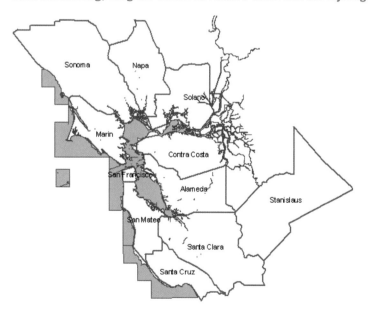

The major hydrographic areas in the BAUA are added to the map display. All hydrographic types in this geodatabase are listed in the table of contents, though not all of them are present in the BAUA.

6   Change the color of the Sea Ocean feature to a deep blue to more closely resemble water.

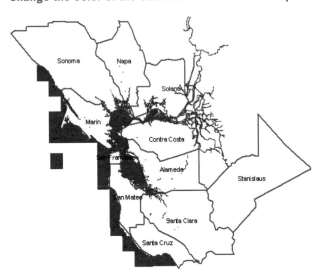

Some hydrologic areas have greater significance to homeland security than others. For example, spillways and dam weirs are critical infrastructure that control the flow and direction of flood waters. These facilities require close surveillance and protection from possible attack.

YOUR TURN

Add the NHDWaterbody, NHDLine, NHDFLowline, and NHDPoint feature classes from ArcCatalog to the Hydrography group layer. Note the location and distribution of reservoirs, pipelines, connectors, tunnels, gates, wells, and water intake/outflow locations, as they are all critical to homeland security planning and operations.

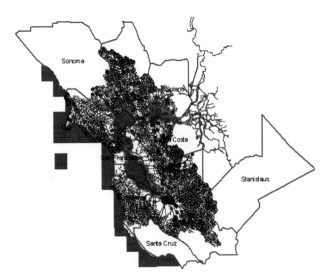

7   Save this map document as **FL_GISHS_C3E4**, with FL being your initials, to **\ESRIPress\GISTHS \MYGISTHS_Work**.

Exercise 3.5
Add structures data to the MEDS map

One of the minimum essential datasets specified by the DHS is the "structures" data. This data consists of significant buildings and facilities that are focal meeting places for large groups of people, or house critical infrastructure heavily depended on for everyday operations. These structures include government offices, public infrastructure, power lines and substations, sports arenas and entertainment centers, transportation terminals, houses of worship, major shopping and office complexes, educational and research campuses, industrial parks, hospitals, and military installations.

In an effort to account for these significant buildings and facilities, DHS has initiated the creation of the National Asset Database (NADB)[24]. This asset inventory is designed to be a comprehensive listing of all national assets that may be targeted for attack by terrorists. The DHS Department of Infrastructure Protection (IP) is charged with collecting this data of critical infrastructure and key resources (CI/KR) that is submitted by each state as it applies for DHS funds. As of January 2006, the classified database contained 77,069 sites, including 17,327 commercial properties such as office buildings, malls, and shopping centers; 12,019 government facilities; 8,402 public health buildings; 7,889 power plants; and 2,963 sites with chemical or hazardous materials.[25]

Due to the classified nature of this database and the diversity of these structures, compiling a comprehensive geospatial data layer of all of these locations entails merging data from a variety of sources. While "Structures" is listed in the NMSS list of data layers, it does not contain any data layers available for download. Specific structure data is more commonly gathered at the local level where land use and community planning requires a detailed inventory of parcels and structures in a specific area.

An alternative starting place to gather this type of data is the U.S. Census Bureau's Topologically Integrated Geographic Encoding and Referencing system, commonly known as TIGER data. This contains a set of landmark datasets in point and polygon format that include a broad range of "distinguishing landscape features marking a site or location."[26]

Download landmark data from the ESRI Census TIGER/Line data Web site

1   Navigate your Internet browser to **www.esri.com/data/data_portals.html**.

A list of data portals is listed here where you can download data for use in a GIS.

2 Click Census 2000 TIGER/Line Data.

3 Click Preview and Download from the Free Download side menu.

Free Download

- Preview and Download

4 Select California from the State drop-down list and click Submit Selection; or click CA on the map.

5 Select Landmark Points from the Select by layer drop-down list, and click Submit Selection.

Select by Layer

Landmark Points

Submit Selection

6 Check the boxes next to each BAUA county, to include:

Alameda
Contra Costa
Marin
Napa (Landmark Points not available)
San Francisco
San Mateo
Santa Clara
Santa Cruz
Solano
Sonoma
Stanislaus

Available counties	File Size
☐ ALL COUNTIES	400.6 KB
☑ Alameda	10.6 KB
☐ Alpine	922.0 bytes
☐ Amador	2.8 KB
☐ Butte	3.4 KB
☐ Calaveras	3.0 KB
☐ Colusa	1.6 KB
☑ Contra Costa	6.3 KB
☐ Del Norte	1.3 KB
☐ El Dorado	5.2 KB
☐ Fresno	11.3 KB
☐ Glenn	1.8 KB
☐ Humboldt	7.2 KB
☐ Imperial	8.2 KB
☐ Inyo	5.4 KB
☐ Kern	13.3 KB
☐ Kings	4.2 KB
☐ Lake	2.0 KB
☐ Lassen	1.4 KB
☐ Los Angeles	64.9 KB
☐ Madera	6.8 KB

7   Click Proceed to Download.

Note: If you have limited Internet connectivity, this file is available locally at \ESRIPress \GISTHS\GISTHS_C3\Downloaded_Data_C3\BAUA_Landmarks. Copy it into your \ESRIPress\GISTHS\MYGISTHS_Work folder before you begin working with it.

8   When the file is ready, click Download File.

9   Click Open to extract the files to **\ESRIPress\GISTHS\MYGISTHS_Work**. You are then required to do a second file extraction of each zip file.

YOUR TURN

Download the BAUA landmark polygon shapefiles from the ESRI Census TIGER/LINE data Web site. Extract them to \ESRIPress\GISTHS\MYGISTHS_Work.

Merge multiple shapefiles

These TIGER shapefiles that are downloaded from the ESRI Census TIGER/Line data Web site have a geographic coordinate spatial reference system that uses longitude and latitude, but cannot display on the current map that already has a defined coordinate system. ArcCatalog allows the user to assign a coordinate system to a shapefile or feature class by either interactively selecting the coordinate system's parameters, or importing them from another shapefile or feature class with the same coordinate system. The National Hydrography Dataset geodatabase also uses the geographic coordinate system spatial reference, so you will import those parameters to the landmark shapefiles. Since there are multiple shapefiles that require importing the coordinate system, you will first merge all of the landmark point shapefiles together so you will only need to import the coordinate system once.

1   Open ArcCatalog if it isn't already opened.

2   Open ArcToolbox from within ArcCatalog.

The ArcToolbox now appears inside of the ArcCatalog window.

3   Expand the Data Management General toolset and double-click Merge.

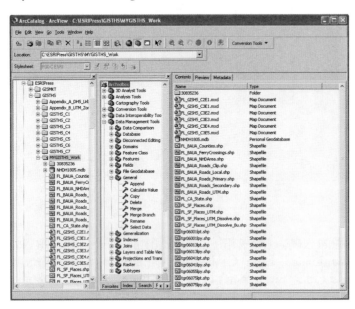

4   Drag all 10 of the landmark point shapefiles from ArcCatalog into the Input Datasets list in the Merge dialogue box.

5   Set the output dataset to: **\ESRIPress\GISTHS\MYGISTHS_Work\FL_BAUA_Landpts_Merge.shp**, with **FL** being your initials.

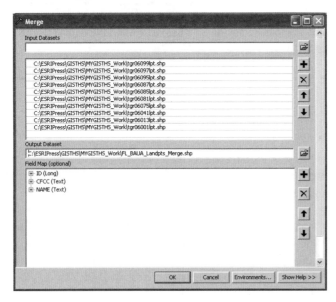

6   Click OK.

7   When the merge is completed, click Close.

The new shapefile now contains all of the landmark points from each of the BAUA counties.

Merge the 11 landmark polygon shapefiles into one shapefile, and save the outset dataset to \ESRIPress\GISTHS\MYGISTHS_Work\FL_BAUA_Landply_Merge.shp, with FL being your initials.

Define landmark shapefile coordinate system

1   In ArcCatalog, right-click FL_BAUA_Landpts_Merge.shp, and select Properties.

2   Click the XY Coordinate System tab.

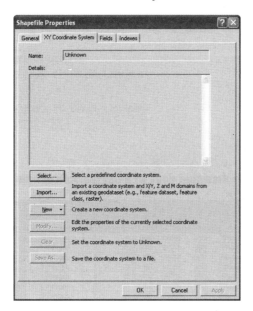

The coordinate system for this shapefile is unknown.

3   Click Import.

EXERCISES

Create and import layer files

When you assign symbology to a data layer, it is only assigned for use in the current data frame. Assigning classification schemes and symbols can take a considerable amount of time if there are numerous unique features in the feature class. To store the assigned symbology of a feature class, it can be saved as a layer file (.lyr) and used across several maps. The symbology can also be imported into a new data layer to expedite the symbol selection process.

1   Right-click FL_BAUA_Landply and select properties.

2   Click the Symbology tab and select Categories and Unique Values.

3   Select DESCRIPTIO from the Value Field drop-down list.

4   Click the Add All Values button.

5   Hold down the Ctrl key and select all water features, to include:

> Bay, estuary, gulf, or sound
> Bay, estuary, gulf, sound, sea, or ocean
> Intermittent lake or pond
> Intermittent stream, river, or wash
> Perennial canal, ditch, or aqueduct
> Perennial lake or pond
> Perennial reservoir
> Perennial stream or river
> Sea or ocean
> Water feature

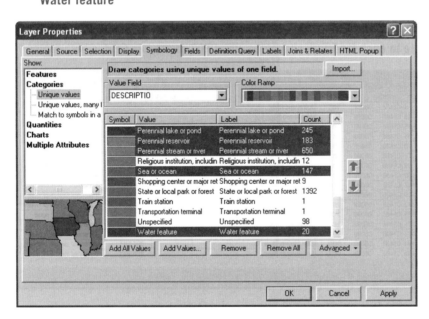

These features are already included in the National Hydrography Database layer so they can be removed from the landmarks dataset.

EXERCISES

6　Click OK.

7　When the merge is completed, click Close.

The new shapefile now contains all of the landmark points from each of the BAUA counties.

YOUR TURN

Merge the 11 landmark polygon shapefiles into one shapefile, and save the outset dataset to \ESRIPress\GISTHS\MYGISTHS_Work\FL_BAUA_Landply_Merge.shp, with FL being your initials.

Define landmark shapefile coordinate system

1　In ArcCatalog, right-click FL_BAUA_Landpts_Merge.shp, and select Properties.

2　Click the XY Coordinate System tab.

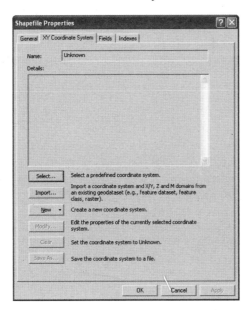

The coordinate system for this shapefile is unknown.

3　Click Import.

EXERCISES

4   Browse for the coordinate system of the NHDPoint data layer in the NHD geodatabase.

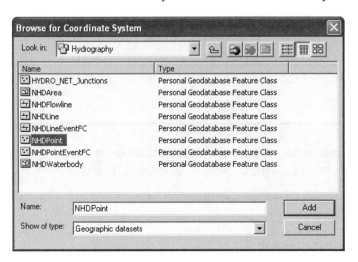

5   Click Add.

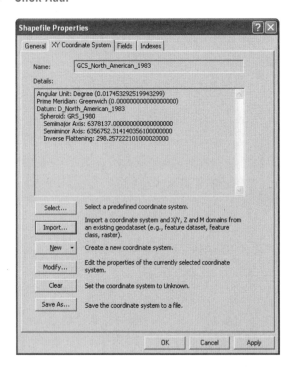

6   Click OK.

The coordinate system parameters are imported into this shapefile.

YOUR **TURN**

Import the NHDPoint coordinate system to the landmark polygon shapefile.

Export landmark shapefiles to data frame coordinate system

1   Return to the ArcMap map document **GISTHS_C3E4** and add FL_BAUA_Landpts_Merge.shp and FL_BAUA_Landply_Merge.shp.

2   Save this map document as **FL_GISHS_C3E5**, with FL being your initials, to **\ESRIPress\GISTHS \MYGISTHS_Work**.

3   Export FL_BAUA_Landpts_Merge.shp to a new shapefile using the same coordinate system as the data frame, and name it **FL_BAUA_Landpts.shp** in the **\ESRIPress\GISTHS\MYGISTHS_Work** folder.

4   Export FL_BAUA_Landply_Merge.shp to a new shapefile using the same coordinate system as the data frame, and name it **FL_BAUA_Landply.shp** in the **\ESRIPress\GISTHS\MYGISTHS_Work** folder.

5   Remove FL_BAUA_Landpts_Merge and FL_BAUA_Landply_Merge from the Table of Contents.

Create new Structures Group layer

1   Create a Structures Group layer in the BAUA MEDS data frame.

2   Drag FL_BAUA_Landpts and FL_BAUA_Landply into the new Structure group layer.

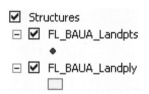

3   Join the cfcc.dbf table to each landmark shapefile using CFCC as the join field in both tables.

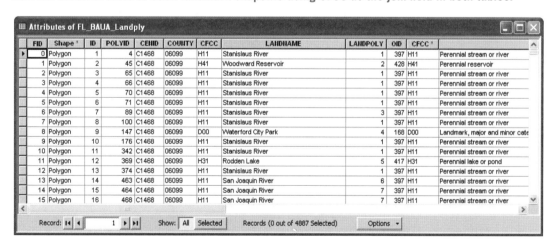

FID	Shape	ID	POLYID	CENID	COUNTY	CFCC	LANDNAME	LANDPOLY	OID	CFCC	
0	Polygon	1	4	C1468	06099	H11	Stanislaus River	1	397	H11	Perennial stream or river
1	Polygon	2	45	C1468	06099	H41	Woodward Reservoir	2	428	H41	Perennial reservoir
2	Polygon	3	65	C1468	06099	H11	Stanislaus River	1	397	H11	Perennial stream or river
3	Polygon	4	66	C1468	06099	H11	Stanislaus River	1	397	H11	Perennial stream or river
4	Polygon	5	70	C1468	06099	H11	Stanislaus River	1	397	H11	Perennial stream or river
5	Polygon	6	71	C1468	06099	H11	Stanislaus River	1	397	H11	Perennial stream or river
6	Polygon	7	89	C1468	06099	H11	Stanislaus River	3	397	H11	Perennial stream or river
7	Polygon	8	100	C1468	06099	H11	Stanislaus River	1	397	H11	Perennial stream or river
8	Polygon	9	147	C1468	06099	D00	Waterford City Park	4	168	D00	Landmark, major and minor cate
9	Polygon	10	176	C1468	06099	H11	Stanislaus River	1	397	H11	Perennial stream or river
10	Polygon	11	342	C1468	06099	H11	Stanislaus River	1	397	H11	Perennial stream or river
11	Polygon	12	369	C1468	06099	H31	Rodden Lake	5	417	H31	Perennial lake or pond
12	Polygon	13	374	C1468	06099	H11	Stanislaus River	1	397	H11	Perennial stream or river
13	Polygon	14	463	C1468	06099	H11	San Joaquin River	6	397	H11	Perennial stream or river
14	Polygon	15	464	C1468	06099	H11	San Joaquin River	7	397	H11	Perennial stream or river
15	Polygon	16	468	C1468	06099	H11	San Joaquin River	7	397	H11	Perennial stream or river

Record: 1   of   Show: All Selected   Records (0 out of 4887 Selected)   Options ▾

The description of each landmark is now included in the attribute tables.

EXERCISES

Create and import layer files

When you assign symbology to a data layer, it is only assigned for use in the current data frame. Assigning classification schemes and symbols can take a considerable amount of time if there are numerous unique features in the feature class. To store the assigned symbology of a feature class, it can be saved as a layer file (.lyr) and used across several maps. The symbology can also be imported into a new data layer to expedite the symbol selection process.

1   Right-click FL_BAUA_Landply and select properties.

2   Click the Symbology tab and select Categories and Unique Values.

3   Select DESCRIPTIO from the Value Field drop-down list.

4   Click the Add All Values button.

5   Hold down the Ctrl key and select all water features, to include:

> Bay, estuary, gulf, or sound
> Bay, estuary, gulf, sound, sea, or ocean
> Intermittent lake or pond
> Intermittent stream, river, or wash
> Perennial canal, ditch, or aqueduct
> Perennial lake or pond
> Perennial reservoir
> Perennial stream or river
> Sea or ocean
> Water feature

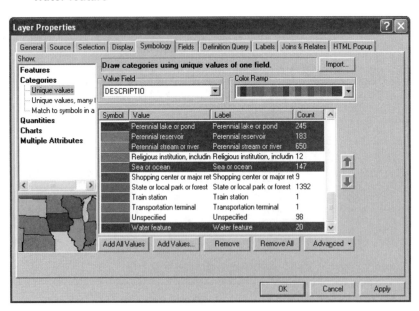

These features are already included in the National Hydrography Database layer so they can be removed from the landmarks dataset.

6   Click Remove.

7   Change the label of the Unspecified field to **Islands**.

8   Double-click the color symbol to the left of the Airport or airfield heading to open the Symbol Selector.

9   Select fill colors and patterns to all of the remaining landmark polygons, as follows:

☑ FL_BAUA_Landply
    cfcc.DESCRIPTIO
    ▨ Airport or airfield
    ▨ Amusement center
    ▨ Campground
    ▨ Cemetery
    ▨ Educational institution, including academy, school, college, and university
    ■ Federal penitentiary, State prison, or prison farm
    ▨ Golf course
    ▨ Government center
    ▨ Hospital
    ▨ Industrial building or industrial park
    ■ Jail or detention center
    ▨ Landmark, major and minor categories unknown
    ▨ Marina
    ▨ Military installation or reservation
    ▨ National park or forest
    ▨ Nursing home, retirement home, or home for the aged
    ▨ Open space
    ▨ Other Federal land
    ▨ Other employment center
    ▨ Religious institution, including church, synagogue, seminary, temple, and mosque
    ▨ Shopping center or major retail center
    ▨ State or local park or forest
    ▨ Train station
    ▨ Transportation terminal
    ▨ Islands

10 Click OK when complete.

The landmark polygon categories are now listed in the table of contents. These symbols represent the most commonly used patterns for these landmarks. To be able to use them again in other map documents as you have classified them here, they can be saved as a layer file (.lyr).

11 Right-click FL_BAUA_Landply and Save as Layer file **FL_BAUA_Landply** in **\ESRIPress\GISTHS \MYGISTHS_Work**, with FL being your initials.

The landmark points features contain many unique categories requiring individual symbol markers to clearly differentiate between them. A layer file has already been created for import to be used to classify and map this data.

12 Right-click FL_BAUA_Landpts and select Properties.

13 Click the Symbology tab and click the Import button.

14 Click the Browse button to navigate the **\ESRIPress\GISTHS\GISTHS_C3 folder**.

15 Select the BAUA_Landpts.lyr file and click Add.

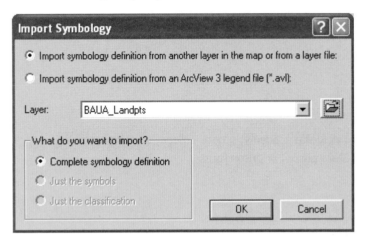

16 Click OK.

17 Select DESCRIPTIO as the value field to use in the imported symbology definition.

EXERCISES

18 Click OK.

The symbols from the layer file are imported into this data layer and used to classify each category of landmark points.

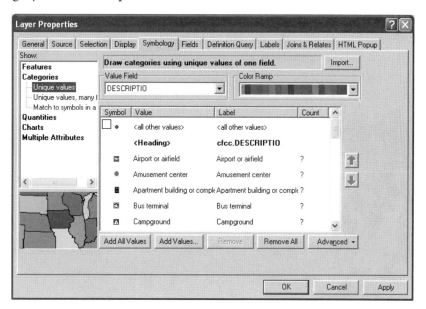

19 Click OK.

☑ FL_BAUA_Landpts
  ◆ <all other values>
      cfcc.DESCRIPTIO
  ▣ Airport or airfield
  ● Amusement center
  ▦ Apartment building or complex
  ▩ Bus terminal
  ▲ Campground
  ⁺⁺ Cemetery
  ◆ Crew of vessel
  ⅃ Educational institution, including academy, school, college, and university
  ⅃ Educational or religious institution
  ↓ Golf course
  ★ Government center
  ⊞ Hospital
  ⌂ Hotel, motel, resort, spa, YMCA, or YWCA
  ◢ Industrial building or industrial park
  ⌐⊤ Jail or detention center
  ● Landmark, major and minor categories unknown

  ⚲ Lookout tower

  ⚓ Marina
  ⌐⁼ Marine terminal
  ▲ Mountain peak, the point of highest elevation of a mountain
  ⌁ National park or forest
  ⊞ Nursing home, retirement home, or home for the aged
  ▦ Office building or office park
  🌲 Open space
  ⌐ Other Federal land
  ◆ Other employment center
  ⅃ Religious institution, including church, synagogue, seminary, temple, and mosque
  ▩ Shopping center or major retail center
  ★ Special purpose landmark
  ▦ State or local park or forest
  ▭ Trailer court or mobile home park
  ⊞ Train station
  ● Transportation terminal

20 Right-click FL_BAUA_Landpts and select Save as Layer File.

21 Save this layer as **FL_BAUA_Landpts**, with FL being your initials, in **\ESRIPress\GISTHS \MYGISTHS_Work**.

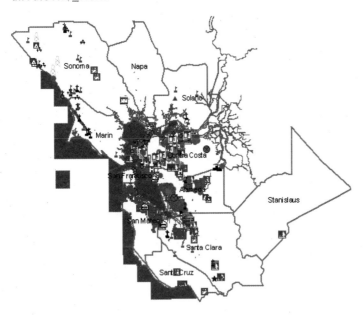

22 Save the map document when complete.

Many of these landmarks are significant buildings and facilities that are focal meeting places for large groups of people, or house critical infrastructure systems that are essential to our daily lives. Identifying their location is crucial to preparing for and responding to an emergency event, be it a natural disaster or terrorist attack. Analysis of a select sample of these landmarks is undertaken in a later chapter within the context of the critical infrastructure program.

Exercise 3.6
Add a raster image layer to the MEDS map

Up to this point, you have worked with only vector data in either a shapefile format, or a feature class in a geodatabase. Another type of data format used in a GIS is the raster data structure. It is used most commonly to display aerial and satellite images, as well as surface models of continuous data representing such features as elevation and climate data.

There are three raster datasets in the MEDS map:

- Orthoimagery
- Land Cover
- Elevation

Orthoimagery is aerial photography where distortion and relief displacement is removed so ground features are displayed in their true planimetrically correct position. Orthoimagery supports making direct measurements of locations, distances, angles, directions, and areas while offering a realistic visualization of the landscape.[27] High-resolution orthoimagery at one-foot resolution, as stipulated for the Urban Area MEDS, enables planners and first responders to visually see geographic features on the ground that may not be mapped, and may also be instrumental in a rescue operation or securing a location in a time of elevated threat or catastrophic event. For example, orthoimagery is used by fire, police, and other emergency response teams to see a picture of a call location. This advantage can decrease response times to on-site rescues and evacuations by as much as 20-30 seconds, a time savings that is agreed by responders to save lives and property[28]. Before and after orthoimagery is critical to recovery efforts for damage detection and discerning the impact of an event, such as the tsunami in the Indian Ocean, or the Ground Zero World Trade Center site.[29]

High-resolution imagery is stored and delivered in often very large graphics files. A complete set of orthoimagery for an area the extent of the BAUA may be many terabytes in size and difficult to manage in our current tutorial dataset. Therefore, you will download a high-resolution orthoimage of a select location of the BAUA that you will use in a later exercise.

Download orthoimagery from the National Map Seamless Server

1 **Open your Web browser and navigate to http://seamless.usgs.gov.**

2 **Click the link View and Download United States Data.**

3 **Click the Zoom to a Point button.**

4 **Type -122.250556, 37.871111 into the dialog box.**

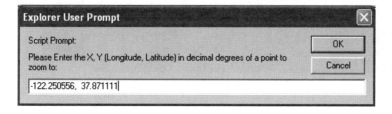

This is the XY longitude and latitude of Memorial Stadium in Berkeley, California.

EXERCISES

5 Click OK.

6 Click off all display layers except the National Atlas States, National Atlas Counties 2001, and GTOP060 Color Shaded Relief layers.

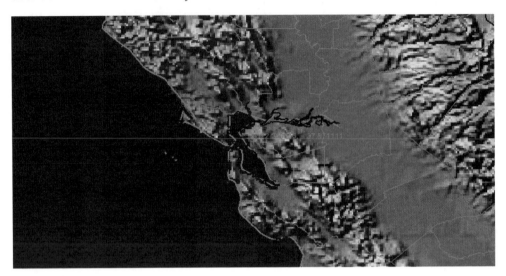

The XY longitude and latitude location is marked on the map display.

7 Use the Zoom In tool ⊕ to zoom in tighter to the BAUA extent.

8 Click to put a check in the box next to the San Francisco-Oakland (Feb 2004) layer under the Orthoimagery group layer.

☑ San Francisco - Oakland (Feb 2004)

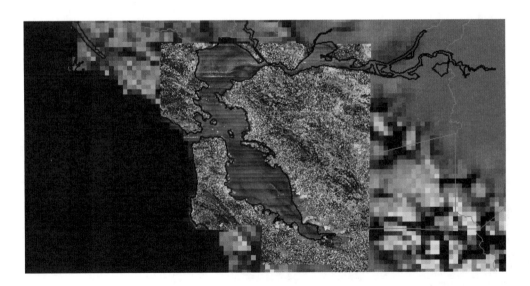

The orthoimagery layer of the San Francisco-Oakland area is added to the map display.

9   Click the layer name in the legend to view the metadata about this layer.

✓ San Francisco -
   Oakland (Feb 2004)

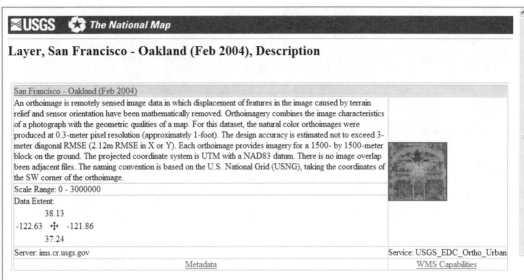

This orthoimagery was produced at 0.3-meter pixel resolution, or approximately 1-foot. This is compliant with the minimum resolution for Urban Area orthoimagery stipulated in the MEDS.

10   Close the metadata window.

11   Use the Zoom In tool to drag a tight box around the XY point marked on the map display.

12 Continue to zoom in until the map display is centered on Memorial Stadium.

13 Click the Downloads tab and click to put a check in the box next to the San Francisco-Oakland (Feb 2004) layer under the Orthoimagery group layer. Be sure there are no other data layers checked in any of the other group layers.

14 Click the Define Download Area button and drag a box around the extent of the map display.

If the box is drawn in green, then this data extent is available for download or ordered on media. If it is drawn in red, then it is too large for download, and is available only on media.

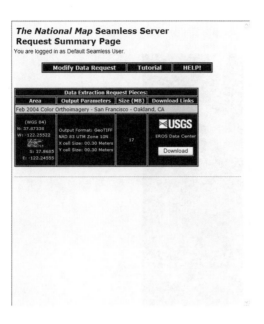

The Request Summary Page appears showing the orthoimagery GeoTIFF file ready for download.

15 Click the Download button.

16 When the file is ready, save and extract it to **\ESRIPress\GISTHS\MYGISTHS_Work**.

Note: If you have limited Internet connectivity, this compressed zip file is available locally at \ESRIPress\GISTHS\GISTHS_C3\Downloaded_Data_C3\BAUA_Orthoimagery. Copy it into your \ESRIPress\GISTHS\MYGISTHS_Work folder before you begin working with it.

17 Close the Internet browser.

Add raster orthoimagery layer to map document

1 Return to the ArcMap map document **GISTHS_C3E5**.

2 Create a new group layer and name it **Orthoimagery**.

3 Add the orthoimage of Memorial Stadium to the Orthoimagery group layer.

4   Turn off the Structures layers and zoom to the extent of the orthoimage layer.

The orthoimagery layer is added to the map display.

5   From the standard toolbar, select 1:100,000 map scale to zoom out in the map display, and turn on the FL_BAUA_Places_UTM data layer.

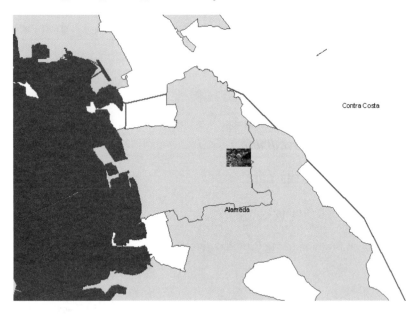

This map scale provides a better spatial context for the location of this site.

6   Save this map document as **FL_GISHS_C3E6**, with FL being your initials, to **\ESRIPress\GISTHS \MYGISTHS_Work**.

EXERCISES

Review raster layer properties

1   Zoom to the extent of the orthoimage layer.

2   Right-click the orthoimagery raster layer and select Properties.

3   Click the General tab and rename layer **Memorial Stadium**.

4   Click the Source tab to review the raster information.

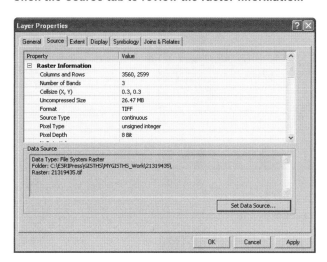

Note the cell size of this orthoimage is 0.3, which means that each pixel on the screen represents one tenth of a meter, or approximately 1 foot on the ground.

5   Scroll down to review the spatial reference of this orthoimage.

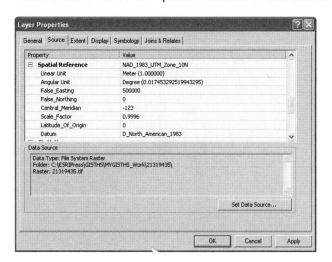

Note that it is projected into NAD_1983_UTM_Zone_10N, and aligns directly with the data already loaded into the map document. Keep in mind that while ArcMap is capable of reprojecting data layers on the fly, it is best to have all of your data layers in a common spatial reference to avoid alignment problems in your map document.

EXERCISES

6   Click the Extent tab to review the extent settings for this layer.

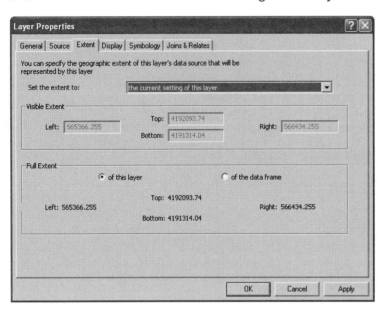

7   Click the Display tab to review the display settings for this layer.

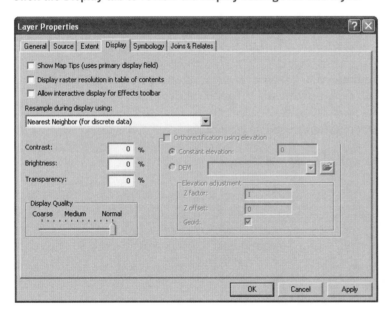

The contrast, brightness, transparency, and display quality properties can be set at this dialog box.

8   Click the Symbology tab to review the symbology settings for this layer.

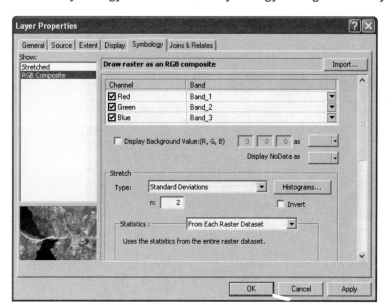

This is a three-band RGB Composite image. Band 1 is shown in the red channel; band 2 in the green channel; and band 3 in the blue channel. All of these bands are captured from the visible light portion of the electromagnetic spectrum, so this image shows landscape detail and color as it appears to the human eye. By manipulating the band, color, contrast, brightness, and data-quality settings of this type of image, you can emphasize particular landscape details of special interest.

Manipulate raster image band properties

1   Click the channels on and off, and click Apply to change the color channels assigned to this composite raster image.

2   Click the down-arrow next to the band, and click Apply to select alternative band assignments to change the band order of this composite raster image.

3   Reset the channel and band settings to the original default setting to return the orthoimage back to its original state.

You can also display only one band of this three-band image at a time.

EXERCISES

4   Click Stretched in the Show: window, and click Apply.

Certain image details may be more evident in one band than in another. For example,

5   Zoom into Band 1 at the stand of trees to the west of the stadium.

6 Change the band selection to Band 2.

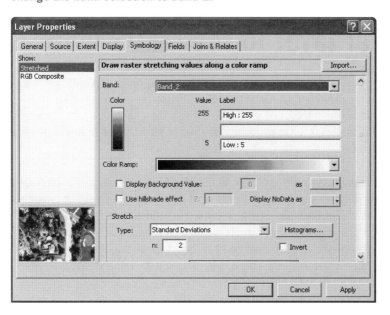

7 Click OK.

Note the enhanced detail in the trees and shadows. This enhancement may be critical to an operation where sharpened detail of landscape features assists in site surveillance.

8 Return the orthoimage back to its original RGB Composite settings.

The MEDS dataset is now equipped with a high-resolution orthoimage of a potential event site in the Bay Area. You will use it in a later chapter in preparation for an upcoming event that requires heightened security and surveillance.

9 Save the map document when complete.

Exercise 3.7
Add Land Cover layer to the MEDS map

The National Land Cover Database of 2001 is a three-product dataset that includes percent tree canopy, percent urban imperviousness, and 16 classes of land cover, all at 30-meter cell resolution from nominal year 2001 Landsat imagery. This database is produced by the Multi-Resolution Land Characteristics (MRLC) Consortium, a group of 13 federal programs in 10 agencies that partner to purchase Landsat imagery and create land cover products for the nation. [30]

The land cover and impervious surface datasets are important to homeland security planning and operations. An inventory of land cover provides planners and operations personnel detailed information about land uses and ground conditions. Knowing the "lay of the land" assists in preventing and mitigating the impacts of a catastrophic event, such as how a plume of toxic gas impacts a region as it passes along a corridor of developed to forested land cover. Though you will only use the land cover layer in this exercise, the percent of urban imperviousness is also important in the event of a flooding emergency. Knowing the proximity to water bodies and the degree to which the land can absorb rainfall during an emergency event assists planners and personnel in managing storm water systems to stave off potential inundation.

Download land cover data from the National Map Seamless Server

1   Open your Web browser and navigate to **http://seamless.usgs.gov/**.

2   Click the link **View and Download United States Data.**

3   Click the **Downloads** tab and click to put a check in the box next to in the **NLCD 2001 Land Cover** layer in the **Land Cover** group layer. Be sure there are no other layers checked.

4.   Click the **Define Download Area by Coordinates** tool in the **Downloads Tools** group box.

5   Click to **Switch to Decimal Degrees.**

6   Enter the same coordinates used when downloading the transportation data:

Note: Your coordinate values may be slightly different.

Top	38.321
Right	-121.412
Bottom	36.979
Left	-122.696

7   Click Add Area and OK or Yes to submit your request.

8   Click Download to download the Land Cover file. Save it to the **\ESRIPress\GISTHS\MYGISTHS_ Work** folder.

> Note: If you have limited Internet connectivity, this compressed zip file is available locally at \ESRIPress\GISTHS\GISTHS_C3\Downloaded_Data_C3\BAUA_ LandCover. Copy it into your \ESRIPress\GISTHS\MYGISTHS_Work folder before you begin working with it.

9   Extract the zip file to \ESRIPress\GISTHS\MYGISTHS_Work.

Add land cover layers to map document

1   Return to the ArcMap map document GISTHS_C3E6.

2   Save this map document as **FL_GISHS_C3E7**, with FL being your initials, to **\ESRIPress\GISTHS \MYGISTHS_Work**.

3   Create a new group layer and name it **Land Cover.**

4   Add the land cover raster dataset to the group layer.

5 Review the spatial reference properties of this raster layer.

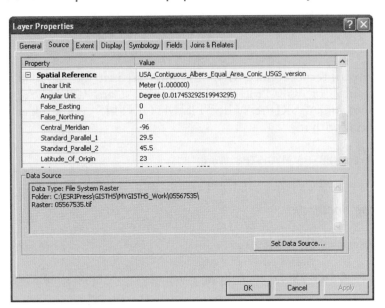

The spatial reference of this raster layer is USA_Contiguous_Albers_Equal_Area_Conic_USGS_version. Keep in mind that while ArcMap is capable of reprojecting data layers on the fly, it is best to have all of your data layers in a common spatial reference to avoid alignment problems in your map document.

6 Close the Layer Properties window.

Export raster data layer

1 Right-click the land cover raster data layer in the Table of Contents and click Export.

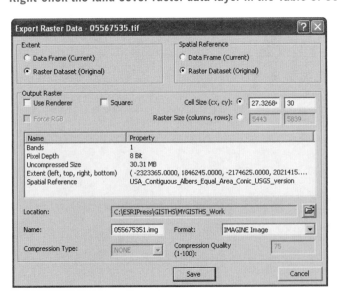

A different export dialog box opens than the one used to export vector data layers.

2   In the Spatial Reference box, click the Data Frame radio button; select GRID format; and name the file **FLBAUALC**, with FL being your initials. Note that GRID layer names can contain no more than eight characters.

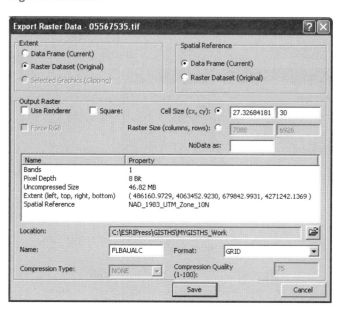

3   Click Save.

4   Click Yes to add the exported data to the map as a layer.

5   Drag **FLBAUALC** into the Land Cover group layer.

6   Remove the original land cover raster data layer.

The new land cover raster layer is drawn to the map display. Conventional land cover colors have already been assigned to the raster values so the display appears close to true color.

Assign labels to raster values and colors

1 Open the Symbology window and click Unique Values to show the list of values and colors assigned to this raster layer.

2 Reselect VALUE as the Value Field to reclassify the color assignments.

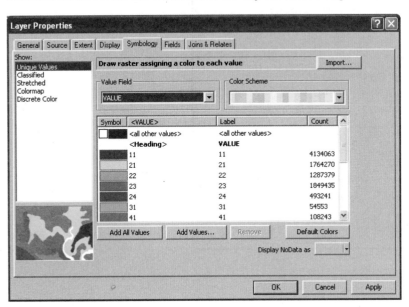

3 Appropriate colors have already been assigned to each value, though the labels do not contain sufficient land cover information. Edit the labels for each value to the NLCD land cover classifications, as follows:

11	Open Water
21	Developed, Open Space
22	Developed, Low Intensity
23	Developed, Medium Intensity
24	Developed, High Intensity
31	Barren Land
41	Deciduous Forest
42	Evergreen Forest
43	Mixed Forest
52	Shrub/Scrub
71	Grassland/Herbaceous
81	Hay/Pasture
82	Cultivated Crops
90	Woody Wetlands
95	Emergent Herbaceous Wetlands

4   Click Remove to remove value 127 from the classification list.

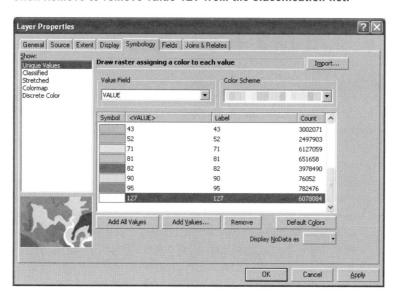

5   Click OK.

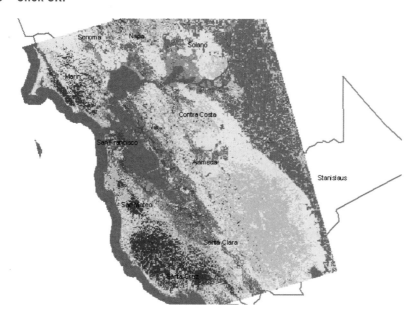

The land cover raster dataset is drawn to the map, displaying the distribution of land cover types throughout the BAUA. This information is helpful to homeland security planners and operations personnel as they prepare for and plan how an emergency event may impact this area.

6   Save the map document when complete.

Exercise 3.8
Add a DEM layer to the MEDS map

A digital elevation model (DEM), also known as a digital terrain model (DTM), is a representation of continuous elevation values over a topographic surface typically used to present terrain relief.[31] DEMs are generated and distributed by the USGS, and are used in the production of 3D graphics displaying terrain slope, aspect (direction of slope), and terrain profiles between selected points.[32]

DEM data is helpful in homeland security planning and operations by providing terrain detail and relief for planning and postevent response and recovery. DEM data is used to identify optimal viewsheds for the placement of surveillance positions, as well as recognizing site access constraints due to terrain impediments. It is also used to orthorectify aerial imagery to remove terrain distortion and relief displacement. Using orthoimagery to create a realistic 3D landscape model helps planners effectively implement protection, response, and recovery plans for the deployment of emergency personnel and resources.

DEM data is available from a variety of sources. The USGS distributes individual DEM files of its 7.5 (1:24,000) quadrangle series through commercial vendors.[33] These datasets are useful for large scale/small area studies. For larger areas where multiple DEM datasets are needed, the National Elevation Dataset (NED) data is available via the NMSS. This data is a raster product designed to provide national elevation data in a seamless form with a consistent datum, elevation unit, and projection. NED has a resolution of one arc-second (approximately 30 meters) for the conterminous United States, Hawaii, and Puerto Rico and a resolution of two arc-seconds for Alaska. (An arc-second represents the distance of latitude or longitude traversed on the earth's surface while traveling one second [1/3600th of a degree.]) Data at the 1/9 arc-second resolution, as stipulated in the MEDS framework, is only available via the NMSS for a few selected areas around the country, but is continuously updated as data becomes available.

Elevation data is stored and delivered in often very large files. A complete set of elevation data for an area the extent of the BAUA is too large to effectively manage in our current tutorial dataset. Therefore, you will download elevation data of a select BAUA location you will use in a later exercise.

Measure spatial extent of DEM area

1   Open map document **GISTHS_C3E7**, if not already opened.

2   Turn off all group layers except Boundaries and Orthoimagery.

3   Zoom to the extent of the Memorial Stadium orthoimage.

4   Change the map scale to 1:22,000.

5   Click the Measure Tool on the Tools toolbar.

6   Click the Choose Units button and set the Area units to Miles.

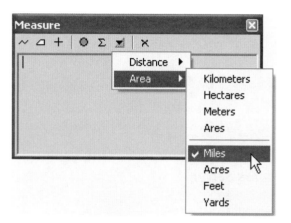

7   Use the Measure An Area tool 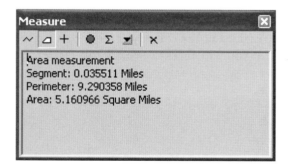 to draw a box around the full extent of the 1:22,000 map display.

Note the measurements displayed in the measure window. The area is approximately 5 square miles. This is an appropriate scale to manage the surveillance and security of an event taking place at Memorial Stadium.

EXERCISES

8   Click along each corner edge of the map display and record the four bounding coordinates of this map extent:

Note: Your coordinate values may be slightly different.

Top	37.884
Right	-122.224
Bottom	37.857
Left	-122.278

Download DEM from the National Map Seamless Server

1   Open your Web browser and navigate to http://seamless.usgs.gov/.

2   Expand the Layer Extent group under the Display tab, and click the Zoom to Full Extent button.

3   Click to check the NED 1/9 Data Coverage layer.   ☑ NED 1/9 arc sec

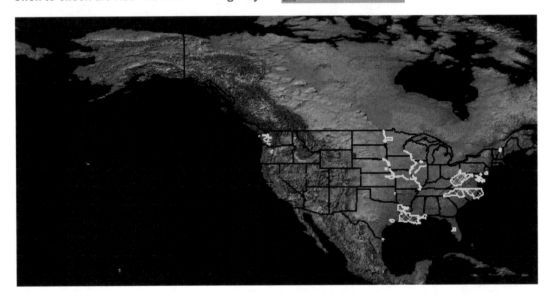

As of this writing, the NED 1/9-arc data is only available in selected areas of the United States. As this DEM data is generated for other regions of the country, it will be added to the NMSS. Since it is not yet available for the BAUA, you will check to see the coverage extent of the next best DEM data.

4 Uncheck the NED 1/9 layer, and click to check the NED 1/3 Data Coverage layer.

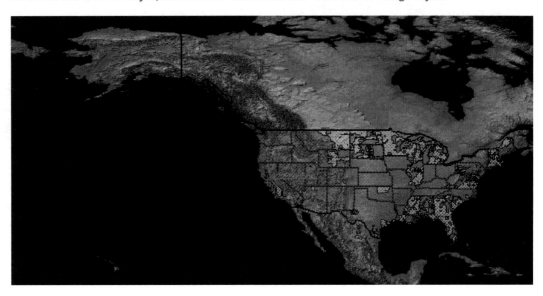

All of the United States is covered by this data layer, though there are two kinds of data symbolized on the map.

5 Zoom to the extent of the BAUA.

6 Click the NED 1/3 Data Coverage layer to view the metadata of this layer.

☑ NED 1/3 arc sec

NED 1/3 arc sec	
[No Description Available]	
Scale Range: Available at all scales.	
Data Extent: 90 -180 ✛ 180 -90	10 meter or better 30 meter source
Server: ims.cr.usgs.gov	Service: USGS_EDC_Elev_NED_3
[No Metadata Specified]	WMS Capabilities

The DEM data for the areas overlaid by the blue hatch pattern is generated from 10 meter or better data resolution. The data in the yellow areas is from 30 meter source data, and not as accurate. Most of the BAUA DEM data is generated from the higher resolution 10 meter or better data.

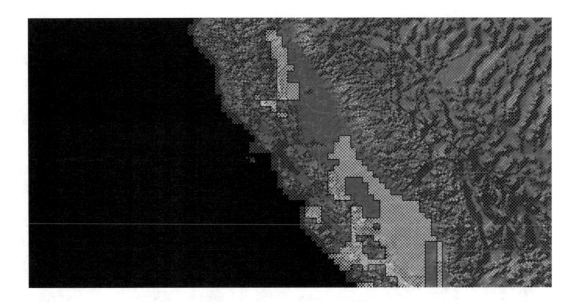

7   Close the metadata window.

8   Click the Download tab, and expand the Elevation group layer.

9   Click to check the 1/3 NED data layer. Be sure no other data layers are checked in this or any other group layer.

10 Click the Define Download Area by Coordinates tool in the Downloads Tools group box.

11 Enter the bounding coordinates of the 1:22,000 spatial extent around Memorial Stadium recorded earlier.

12 Click Add Area and OK or Yes to submit your request.

13 Download and extract the elevation data zip file to **\ESRIPress\GISTHS\MYGISTHS_Work**.

Note: If you have limited Internet connectivity, this file is available locally at \ESRIPress \GISTHS\GISTHS_C3\Downloaded_Data_C3\BAUA_DEM. Copy it into your \ESRIPress\GISTHS\MYGISTHS_Work folder before you begin working with it.

Add elevation data to the MEDS map

1   Return to the ArcMap map document GISTHS_C3E7.

2   Create a new group layer and name it **Elevation**.

3   Add the elevation raster dataset to the Elevation group layer and zoom to its extent.

EXERCISES

4   Review the spatial reference properties of this raster layer.

The spatial reference of this raster layer is GCS_North_American_1983. Keep in mind that while ArcMap is capable of reprojecting data layers on the fly, it is best to have all of your data layers in a common spatial reference to avoid alignment problems in your map document.

5   Close the Layer Properties window.

6   Open ArcToolbox and navigate to Data Management Tools > Projections and Transformations > Raster > Project Raster.

7   Set the input raster as your downloaded NED raster.

8   Set the output file to **FL_BAUA_DEM**, with FL being your initials, in your **\ESRIPress\GISTHS \MYGISTHS_Work** folder.

9   To set the output coordinate system, press Select > Projected Coordinate System > UTM > NAD 1983 > NAD 1983 UTM Zone 10.

10   Select BILINEAR as the Resampling Technique.

11   Accept the default output cell size.

12   Click OK.

13   When the projection process is complete, click Close.

14   Remove the original downloaded DEM data layer from the Table of Contents.

15 Drag the newly projected DEM layer into the Elevation group.

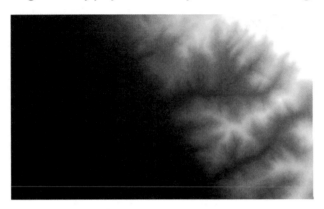

The elevation raster dataset is drawn to the map display. It is classified using a graduated grayscale ramp with lower elevation areas shown in dark shades, and higher elevation areas shown in lighter shades.

16 Turn on the NHDFLowline data layer in the Hydrography group layer and drag it above the FL_ BAUA_DEM raster data layer in the Table of Contents.

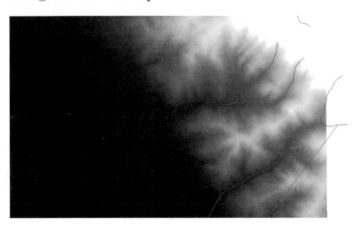

These data layers show the relationship between the flow of water features and the contour of the landscape.

ArcMap Spatial Analyst extension provides additional functionality and processing tools to generate contour, slope, aspect, hillshade, viewshed, and cut and fill data layers from elevation models. You will use some of these tools in later chapters of this tutorial to prepare a DEM for use in a homeland security scenario that requires heightened surveillance within proximity to an event site.

17 Save this map document as **FL_GISHS_C3E8**, with FL being your initials, to **\ESRIPress\GISTHS \MYGISTHS_Work**.

Exercise 3.9
Add Geographic Names to the MEDS map

The final piece of the MEDS map remaining is the Geographic Names data layer. The Geographic Names Information System (GNIS), developed by the U.S. Geological Survey in cooperation with the U.S. Board on Geographic Names, contains information about physical and cultural geographic features in the United States and associated areas, both current and historical (not including roads and highways). The database holds the federally recognized name of each feature and defines the location of the feature by state, county, USGS topographic map, and geographic coordinates.[34]

This information is critical to homeland security planning and operations, as it is the source for applying geographic names to federal maps. It is from this database that the names are assigned to geographic features in the National Map. A single names database provides homeland security planners and operations personnel with a reliable, consistent, and common feature identification system for providing protection to critical infrastructure and for identifying landmarks in a response search and rescue operation.

Download GNIS data from the U.S. Board on Geographic Names Web site

1   Navigate your Internet browser to **http://geonames.usgs.gov/domestic/index.html**.

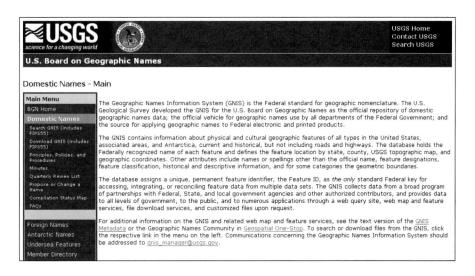

2   Click Download GNIS (includes FIPS55) on the left sidebar of the Web page.

3   Select California from the State drop-down list.

4   Save and extract **CA_DECI.zip** to **\ESRIPress\GISTHS\MYGISTHS_Work**.

Note: If you have limited Internet connectivity, this compressed zip file is available locally at \ESRIPress\GISTHS\GISTHS_C3\Downloaded_Data_C3\BAUA_GeogNames. Copy it into your \ESRIPress\GISTHS\MYGISTHS_Work folder before you begin working with it.

This file is a tabular .txt file. ArcMap is capable of loading a .txt file directly into a map document, but the schema.ini file that defines the text format must be modified before it is importable.

Modify schema.ini file

1   Open ArcCatalog and navigate to **\ESRIPress\GISTHS\MYGISTHS_Work**. If it is already opened, hit F5 to refresh the window.

2   Click CA_DECI.txt in the left panel and click the Preview tab in the right panel. This table cannot be displayed as a table until the schema.ini file is modified to read its format.

3   Open an Explorer window and navigate to the **\ESRIPress\GISTHS\MYGISTHS_Work** folder.

4   Double click schema.ini to open it in Notepad.

5   Change Format=CSVDelimited to Format=Delimited(I). Note that this is the "pipe" symbol that is the shift of the backward slash in the keyboard.

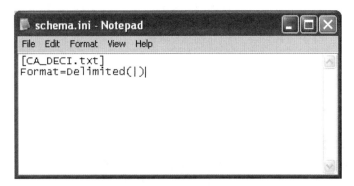

6   Save the file and exit Notepad.

7   Return to ArcCatalog and click on **\ESRIPress\GISTHS\MYGISTHS_Work**. Refresh the screen to see the CA_DECI.txt file now displayed in the Preview window.

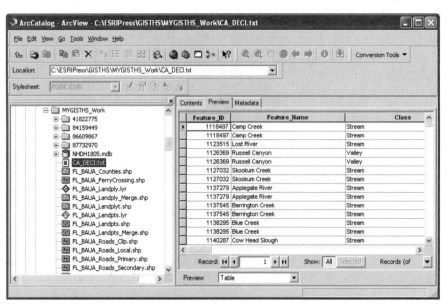

EXERCISES

Create a Geographic Names feature class from XY Table

Now that you have a table of the geographic names, ArcMap enables you to map these features by their recorded x,y coordinates.

1 In ArcCatalog, right-click CA_DECI.txt, and select Create Feature Class and From XY Table.

2 Select Primary_lon_DEC from the X Field drop-down list.

3 Select Primary_lat_DEC from the Y Field drop-down list.

4 Edit the Coordinate System of Input Coordinates to import the same coordinate system set for the NHDPoint layer used earlier.

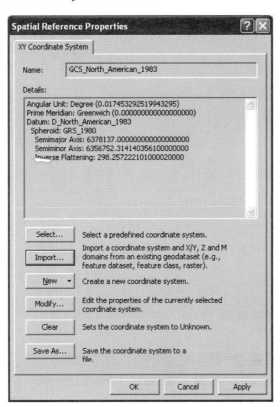

5 Click OK.

6 Save the output as **FL_CA_GNIS.shp**, with FL being your initials, to **\ESRIPress\GISTHS\MYGISTHS_ Work**.

7 Click OK.

Add GNIS shapefile to map document

1   Return to the ArcMap map document FL_GISHS_C3E8.

2   Create a new group layer and name it **Geographic Names.**

3   Add **FL_CA_GNIS.shp** it to the Geographic Names group layer.

All of the geographic names records in California are now plotted as points on the map.

4   Right-click FL_CA_GNIS to open the attribute table.

5   Click ▶| at the lower left of the table window to go to the last record in the database.

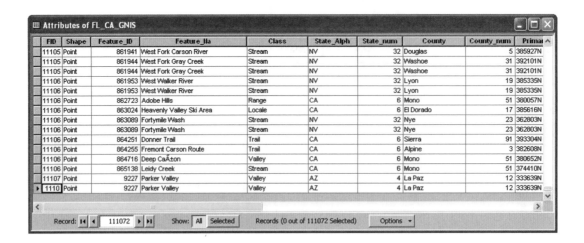

This is a very large database of 111,072 records. We will select only those records that are in the counties of the BAUA.

6   Close the attribute table and save this map document as **FL_GISHS_C3E9**, with FL being your initials, to **\ESRIPress\GISTHS\MYGISTHS_Work**.

Select by attributes

1   From the Main Menu, click Selection, and Select by Attributes.

2   Select the 11 BAUA counties as follows:

[County] = 'Alameda' OR [County] = 'Contra Costa' OR [County] = 'Marin' OR [County] = 'Napa' OR
[County] = 'San Francisco' OR [County] = 'San Mateo' OR [County] = 'Santa Clara' OR
[County] = 'Santa Cruz' OR [County] = 'Solano' OR [County] = 'Sonoma' OR [County] = 'Stanislaus'

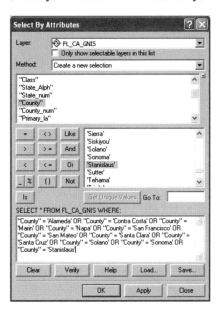

3   Click Apply and Close.

4   Open the attribute table and click the Selected button to view only the records in the BAUA counties.

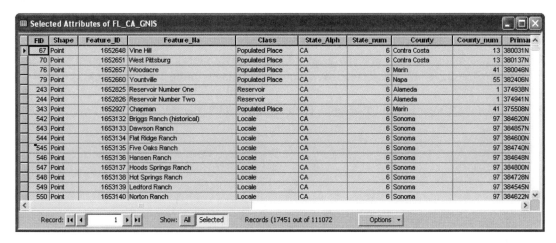

There are 17,451 records out of the total 111,072 records selected from this database.

5   Close the attribute table.

YOUR TURN

Some of the geographic names points are located outside of the actual BAUA boundary. Select by location only the geographic names points that are completely within the BAUA boundary. Export them to a shapefile named FL_BAUA_GNIS.shp, with FL being your initials, in the ESRIPress \GISTHS\MYGISTHS_Work folder. Use the coordinate system of the data frame to convert this layer to the coordinate system common to the other MEDS data layer.

6   Drag the new shapefile into the Geographic Names group layer, and remove FL_CA_GNIS from the Table of Contents.

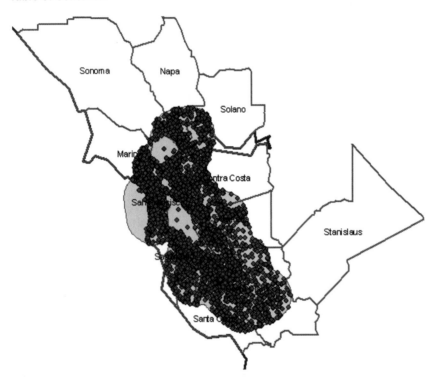

7   Since this feature is used to identify geographic names rather than class or type of feature, change the symbol for these features to a common symbol: **Circle 1 Size 5 Black**.

8   Set the label properties to **Label Field: Feature_Na, Ariel 8, Burnt Umber**.

9 Because this data layer is so dense, it is best to set the label scale range not to show labels when zoomed out greater than 1:24,000. Individual features can be identified using the Identify button.

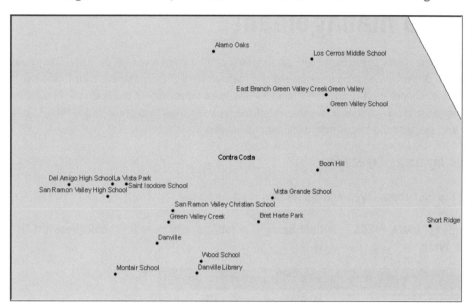

This data layer provides a single consistent source database to identify landscape features in the community. A single names database provides homeland security planners and operations personnel with a reliable, consistent, and common feature identification system for providing protection to critical infrastructure and for identifying landmarks in a response search and rescue operation.

10 Save the map document when complete.

Exercise 3.10
MEDS data management

Your MEDS map document now consists of eight group layers, one for each essential data set. Saving each of these group layers as a layer file enables easy access and management of the data layers for future use. Like layer files of individual shapefiles, the layer files store the assigned symbology of the features. In addition, when you save the entire group layer as a layer file, all of the layers in the group are stored together, so that they are quickly and easily added to any map document as needed.

Save all group layers as layer files

1 **Right-click the first group layer, and select Save as a Layer file.**

2 **Add the prefix FL_BAUA_MEDS_ , with FL being your initials, and save it to \ESRIPress\GISTHS \MYGISTHS_Work.**

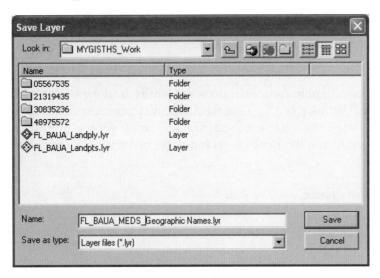

3 **Save all remaining group layers as layer files.**

FL_BAUA_MEDS_Boundaries.lyr	Layer
FL_BAUA_MEDS_Elevation.lyr	Layer
FL_BAUA_MEDS_Geographic Names.lyr	Layer
FL_BAUA_MEDS_Hydrography.lyr	Layer
FL_BAUA_MEDS_LandCover.lyr	Layer
FL_BAUA_MEDS_Orthoimagery.lyr	Layer
FL_BAUA_MEDS_Structures.lyr	Layer
FL_BAUA_MEDS_Transportation.lyr	Layer

You can now add these group layers to any map document as needed. As long as these layer files are stored along with the shapefiles and feature classes that they reference, they will load all the features in the group with the symbology that was assigned in this chapter.

4 **Exit ArcMap and ArcCatalog.**

Summary

You have just completed the first critical step of developing an effective geospatial model for homeland security. In this chapter, you reviewed and built the components of the Minimal Essential Data Sets defined by DHS. The MEDS include data from various Internet portals and are prepared in a range of formats, including vector shapefiles, raster datasets, tables, and geodatabases. Each of the databases was downloaded and processed for inclusion into the MEDS database using a full suite of ArcMap geoprocessing tools to include select, buffer, merge, import, export, clip, and add x,y coordinate data as a data layer. Each layer in the database was then saved as a layer file to enable full use of the data attributes in future homeland security GIS planning and operations scenarios. This MEDS data layer configuration is an effective standardized file and directory structure that will enable interoperable and efficient geospatial data management and analysis in times of crisis.

Having a comprehensive and interoperable geospatial database ready for homeland security planning and operations enables emergency personnel to quickly access the data needed to stave off, mitigate, or respond to crisis situations. Knowing where people and places are in a community helps in pre-event planning and, as an event unfolds, quickens response times to those in need.

In later exercises, you will use some of this data that you built for the BAUA to address the prevention, protection, response, and recovery of communities in crisis scenarios.

Notes

1. WordNet® 3.0, © 2006 by Princeton University.

2. http://www.eomonline.com/Archives/Jan04/Vernon.html

3. FY 2006 Homeland Security Grant Program, December 2, 2005, Appendix H, P. H-3

4. National Geospatial-Intelligence Agency, *Geospatial Intelligence Standards: Enabling a Common Vision*, November 2006, p.4

5. Tri-Valley, etc...

6. National Geospatial—Intelligence Agency, *Geospatial Intelligence Standards: Enabling a Common Vision*, November 2006, p.2

7. U.S. Department of Homeland Security, *State and Local Fusion Centers*, www.dhs.gov/xinfoshare/programs/gc_1156877184684.shtm

8. Ibid.

9. J. Dangermond PowerPoint Slides, AAG, April 19, 2007, Association of American Geographers Conference, San Francisco, CA.

10. USGS, http://pubs.usgs.gov/of/2005/1379/of2005-1379.doc

11. http://nationalmap.gov

12. http://nmviewogc.cr.usgs.gov/viewer.htm

13. *FY 2007 Homeland Security Grant Program: Program Guidance and Application Kit*, U.S. Department of Homeland Security, Office of Grants and Training, January 5, 2007, p. 47

14. http://www.dhs.gov/xnews/releases/pr_1168010425128.shtm

15. *FY 2006 Homeland Security Grant Program*, December 2, 2005, p. 78

16. *DHS Introduces Risk-based Formula for Urban Areas Security Initiative Grants, Attachment A: List of Fiscal Year 2006 Urban Areas Security Initiative Eligible Applicants*, Press Release 01/03/06, http://www.dhs.gov/xnews/releases/press_release_0824.shtm

17. *FY 2007 Homeland Security Grant Program: Program Guidance and Application Kit*, U.S. Department of Homeland Security, Office of Grants and Training, January 5, 2007, p.3

18. Bay Area Regional GIS Council, http://base.bargc.org

19. Bay Area Regional Homeland Security Data Server (BAR_HSDS) Critical Infrastructure Protection Initiative (CIPI 3), August 20, 2003, http://www.baama.org/bargc/documents/BAR-HSDS_CIPI3_v19.pdf, p.1

20. http://nhd.usgs.gov/

21. ArcGIS Desktop Help

22. These enhancements are available with the complete ArcGIS ArcINFO GIS product. Geodatabase structure and storage capabilities are available with the ArcGIS ArcView GIS product.

23. These features are available with the complete ArcGIS ArcINFO GIS product.

24. *Progress in Developing the National Asset Database*, U.S. Department of Homeland Security, Office of Inspections and Special Reviews, June 20, 2006

25. Ibid., p.5.

26. landmark. (n.d.). *Dictionary.com Unabridged* (v 1.1). Retrieved April 12, 2007, from Dictionary.com Web site: http://dictionary.reference.com/browse/landmark

27. http://www.grandriver.ca/index/document.cfm?sec=9&sub1=0&sub2=0

28. Oblique Imagery & Geo-Referenced GIS Data Pilot Project for Homeland Security, Atlanta Regional Commission, August 2005, p.13.

29. See http://www.globexplorer.com/disasterimages for a sample of before and after aerial images of worldwide disaster locations.

30. *PE&RS*, the journal of the American Society for Photogrammetric Engineering & Remote Sensing, April 2007, Volume 73, Number 4, p. 335.

31. ArcGIS Desktop Help

32. http://rockyweb.cr.usgs.gov/elevation/dpi_dem.html#document

33. http://www.gisdatadepot.com, http://www.mapmart.com, http://www.atdi-us.com

34. http://nhd.usgs.gov/gnis.html

Chapter 4:

Designing map layouts for homeland security planning and operations—*Prevent*

DHS National Planning Scenario 5: Chemical attack— *blister agent*

MISSION AREA: *PREVENT*
DHS Target Capability: *Information sharing and collaboration*
DHS Task: *Pre.A.5 3 Disseminating indications and warnings*

MISSION AREA: *PROTECT*
Incident-site and EOC actions to dispatch, detect, and assess, predict, monitor, and sample hazard material.

MISSION AREA: *RESPOND*
Alerts, activation and notification, traffic and access control, protection of special populations, resource support, requests for assistance, and public information.

MISSION AREA: *RECOVER*
Decontamination of immediate concentrated and distant spot contamination; disposal of contaminated waste; environmental testing; and public provision.

4.1 Prepare scenario map
Load data layers
Select by Location (within a distance of)

4.2 Prepare warning report
Join table to an Excel file
Generate report

4.3 Build warning map layout
Choose a pre-built layout template
Create a custom layout for multiple maps
Add multiple data frames to a map layout
Add a callout box to a map layout
Insert map layout elements
Save a custom map template to My Templates
Use a custom map template
Add a report to the map layout

4.4 GIS outputs
Print report-sized layout
Print large poster-sized plot
Export layouts
Other outputs

Review *Prevent* Mission Area planning scenario
Prepare warning report
Build warning map layout
Create a custom map template
Print and export GIS outputs

Chapter 4

Designing map layouts for homeland security planning and operations: *Prevent*

The next four chapters focus on using GIS in the homeland security mission areas of prevention, protection, response, and recovery. While information collection, or in this case, compiling the MEDS data, is critical to prepare a region for an emergency, analyzing MEDS data within the context of an impending event turns geospatial information into meaningful geointelligence useful in all mission areas. This geointelligence helps homeland security planners and operations personnel recognize potentially threatening situations, enabling them to mitigate impacts or even prevent the event from occurring.

DHS defines *prevent* as the deterrence of all potential terrorists from attacking America, detecting terrorists before they strike, preventing them and their instruments of terror from entering our country, and taking decisive action to eliminate the threat they pose.[1] The *Prevent* Mission Area centers upon preventing threats to homeland security by strengthening pre-event planning and heightening situational awareness. Within this mission area, the Target Capabilities List (TCL) identifies *information sharing and collaboration* as one of the critical capabilities to effectively combat potential threats to homeland security.[2] The Universal Task List (UTL) includes a set of tasks within the context of this target capability to detect threats, control access, and eliminate threats to homeland security.[3] *Disseminating indications and warnings* is one of a set of vital tasks identified in this task list to detect threats to national security.[4]

In this chapter, you will address the prevention of a potential chemical attack, as illustrated in the National Planning Scenario 5: Chemical attack—blister agent. Data layers critical to the prevention of this attack are analyzed to produce a series of reports and map layouts to disseminate threat indications and warnings to potential sites of origin.

Chapter 4: Data dictionary

Layer	Type	Layer Description	Attribute	Description
US_States.FIPS.xls	Excel File	State FIPS Designations	FIPS	Federal Information Processing Standards
			STATE	State name
NA_Airports_DAFIF	Shapefile	Digital Aeronautical Flight Information File	STATE_PROV	State or province name
			NAME	Airport name
			ARPT_INDENT	Airport identification number
USA_Interstates	Shapefile	USA Interstate highway polylines		
USA_States	Shapefile	USA State polygons	STATE_NAME	State name

Exercise 4.1
Prepare scenario map

DHS National Planning Scenario 5:
Chemical attack—blister agent

DHS Target Capability:
Information sharing and collaboration

DHS Task:
Pre.A.5 3 Disseminating indications and warnings

DHS National Planning Scenario 5: Chemical attack—blister agent is described as follows:

Agent Yellow, which is a mixture of the blister agents sulfur mustard and lewisite, is a liquid with a garlic-like odor. Individuals who breathe this mixture may experience damage to the respiratory system. Contact with the skin or eye can result in serious burns. Lewisite or mustard-lewisite also can cause damage to bone marrow and blood vessels. Exposure to high levels may be fatal.

In this scenario, the Universal Adversary (UA) uses a light aircraft to spray chemical Agent Yellow into a packed college football stadium. The agent directly contaminates the stadium and the immediate surrounding area and generates a downwind vapor hazard. The attack causes a large number of casualties that require urgent and long-term medical treatment, but few immediate fatalities occur.[5]

The scenario profile includes a synopsis of how the *Prevent* Mission Area is to be activated as a response to this threat:

The ability to prevent the attack is contingent on the prevention of chemical warfare material (CWM) importation, weapon assembly, plane and pilot acquisition, and site reconnaissance. Deterrence measures must be taken by visibly increasing security and apprehension likelihood at the site before and during the attack. Depleting overseas stockpiles of Mustard and precursor agents would also aid in preventing such an attack.[6]

Given the type of aircraft capable of carrying out this mission, planners can identify possible points of origin. The UA is expected to use an aircraft similar to five or six Cessna Aircraft from the 2003 model year, single engine, private planes that have a one-way range of 700 to 800 miles.[7]

The objective of this exercise is to identify and inform possible airport sites beyond the immediate area of the stadium where the plane and pilot may originate. Once identified, maps and reports are generated to alert airport personnel to the possibility of this threat. They can step up surveillance to detect suspicious activity at their airfield, such as the transportation and storage of a commercial rental trailer near the airpark, or the installation of a new spray system onto a newly acquired airplane.

The scenario map is to contain data layers that set the location of the stadium within the BAUA. These data layers are available in the BAUA MEDS dataset, and are to be supplemented by an additional data layer of airports located outside of the immediate area.

Load data layers

1   Open a new ArcMap document.

2   Add the **FL_BAUA_MEDS_**Boundaries, Orthoimagery, and Structures group layers from the
    \ESRIPress\GISTHS\MYGISTHS_Work folder.

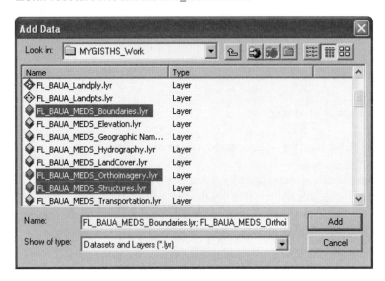

3   Turn off all layers except the **FL_BAUA_Counties**, Memorial Stadium, and **FL_BAUA_Landply**
    data layers.

4   Drag the orthoimagery group layer to the bottom of the Table of Contents.

5   Zoom to the extent of the Memorial Stadium orthoimagery layer.

6   Turn off the Labels for the **FL_BAUA_Counties** data layer.

7   Adjust the transparency of the **FL_BAUA_Landply** data layer to 50%.

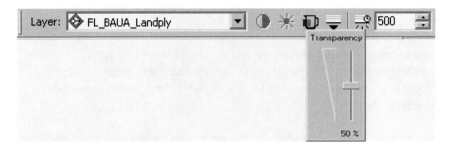

A single polygon encompasses this sports facility.

8   Add additional airports data layer, **NA_Airports_DAFIF.shp**, from **\ESRIPress\GISTHS\GISTHS_C4**, and zoom to layer.

This data layer is a complete database of North American airports. It provides the names and identifier codes for all airports on the continent. Note that the map display is skewed because this map projection is best used for a larger scale map than the full continental extent.

9   Open the Attributes of NA_Airports_DAFIF table to view the contents of this database.

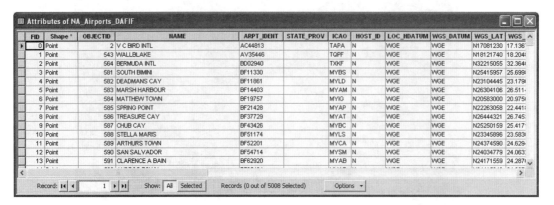

10   Close the attribute table when you have finished reviewing the database contents.

11   Save the ArcMap map document as **FL_GISTHS_C4E1.mxd**, with FL being your initials, to the **\ESRIPress\GISTHS\MYGISTHS_Work** folder.

Select by Location (within a distance of)

The origin of this potential threat is from an airport within an 800-mile range of the target site. This range assumes a one-way flight, with the possibility that the pilot will not make a return trip back to the originating airfield.

1   Zoom to extent of the Memorial Stadium orthoimagery layer.

2   Use the Select Features tool ![cursor icon] to click on the polygon over Memorial Stadium.

3   Select by location features from **NA_Airports_DAFIF** that are within an 800-mile distance of the
    selected **FL_BAUA_Landply** Memorial Stadium site.

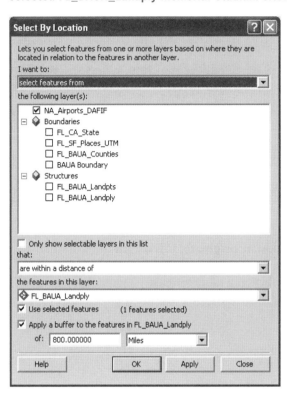

4   Click OK.

5   Zoom to the extent of **NA_Airports_DAFIF**.

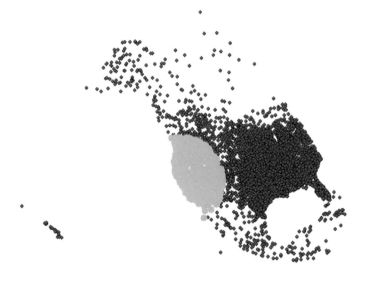

All of the airports within an 800-mile distance of Memorial Stadium are selected.

6   Open the Attributes of **NA_Airports_DAFIF.shp** table.

EXERCISES

7   Click the Selected button to view only the 735 out of 5,008 airports within 800 miles of
     Memorial Stadium.

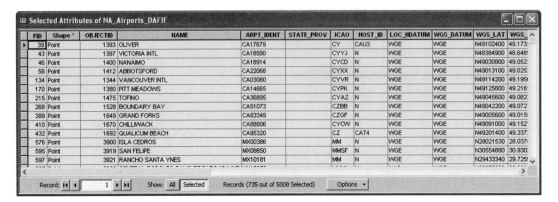

These are the airports where a potential attack may originate. It is useful to have this
information in a GIS to show the spatial distribution of these airports, and to have the
detailed attribute information about each facility in a table for easy viewing. However, this
information must be reported to personnel at these locations to alert them to the possibility
that this situation may arise.

8   Close the Attributes of NA_Airports_DAFIF table.

9   Uncheck the Structures group layer in the Table of Contents.

Exercise 4.2
Prepare warning report

ArcMap has an integrated report writer to create tabular reports from the attribute tables of a data layer. In this scenario, you have selected all of the airports within an 800-mile radius of a potential threat site. You will now generate a warnings report to be disseminated to all airports indicated in this report.

Join table to an Excel file

The Attributes of NA_Airports_DAFIF table includes a numeric FIPS state designation for each airport. You will add an Excel spreadsheet that contains the state name and abbreviation for all 50 states and the District of Columbia so that you can include this information in the warnings report.

1   **Click the Add Data button and navigate to the \ESRIPress\GISTHS\GISTHS_C4 folder.**

2   **Double-click US_States_FIPS.xls.**

3   **Double-click WS_US_States_FIPS$ to add this table to the map document.**

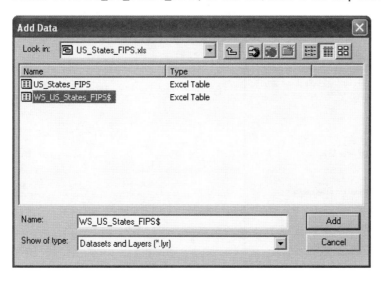

Excel workbooks are imported into ArcMap just as any other tabular data source is, with one difference. Since there can be multiple worksheets in a workbook file, ArcMap allows you to add an individual worksheet to a map document.

4 Open the Attributes of WS_US_States_FIPS$ to view its contents.

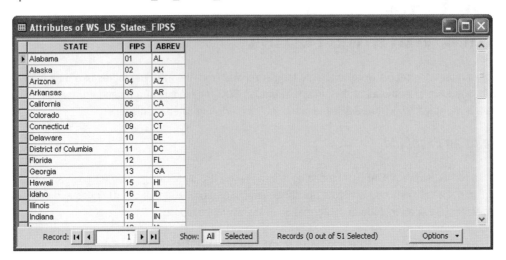

5 Close the Attributes of WS_US_States_FIPS$ table.

6 Join this Excel table to the NA_Airports_DAFIF data layer, using STATE_PROV as the field in the layer that the join will be based on, and FIPS as the field in the table to base the join on.

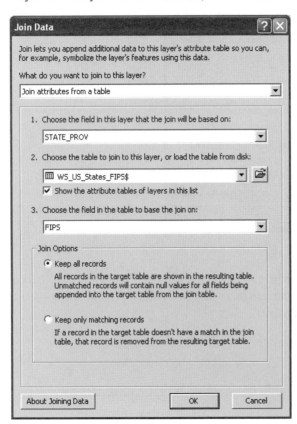

7 Open Attributes of NA_Airports_DAFIF.

8 Click Selected in the NA_Airports_DAFIF table and scroll down and to the far right of the table to see the state designations for the selected records.

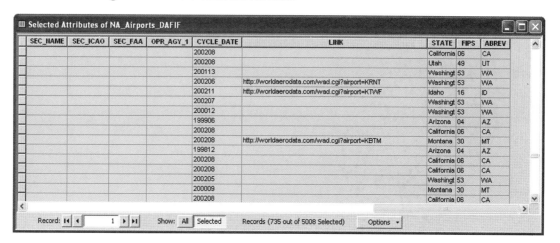

9 Close the Attributes of NA_Airports_DAFIF table.

10 Export the selected features of this data layer to a new shapefile in **\ESRIPress\GISTHS \MYGISTHS_Work** using the coordinate system of the data frame, and name it **FL_Airports_800m. shp**, with FL being your initials.

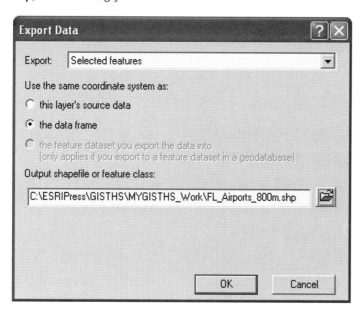

11 Click Yes to add it to the map.

12 Remove NA_Airports_DAFIF.

13   Turn on **FL_CA_State** and zoom to the extent of the **FL_Airports_800m** data layer.

Only those airports within an 800-mile range from the Memorial Stadium site are now shown on the map.

14   Save the ArcMap map document as **FL_GISTHS_C4E2.mxd**, with FL being your initials, to the **\ESRIPress\GISTHS\MYGISTHS_Work** folder.

Generate report

1   From the Main Menu, click Tools, Reports, Create Report.

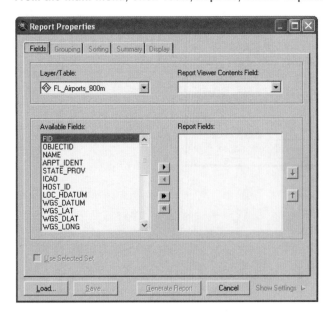

The Report Properties window opens.

2   In the Layer/Table box, select FL_Airports_800m from the drop-down list.

3   In the Available Fields box, double-click NAME, ARPT_IDENT, and STATE (not STATE_PROV).

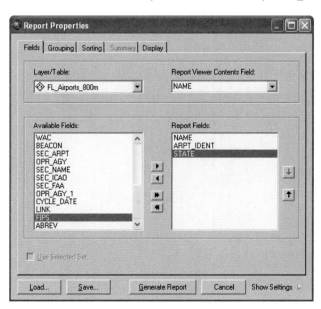

4   Click the Grouping tab and double-click STATE.

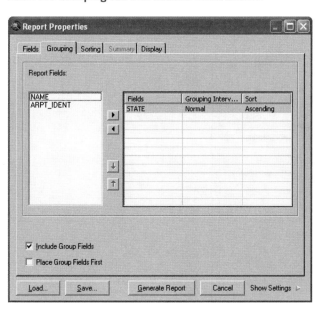

5   Click the Sorting tab and click the Sort field for NAME and select Ascending from the drop-down list.

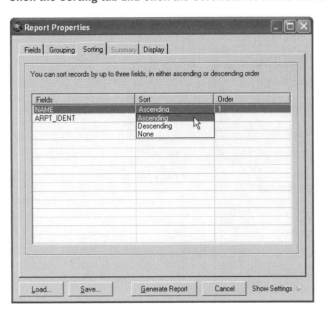

6   Click the Display tab, and click to check on the box to the left of the Title item. Change the Text property from Report Title to Airport Warning Report.

Change the font to Arial Bold 16 by clicking the ⬚...⬚ box and selecting the font from the pop-up window.

7   Click to put a check in the box to the left of the Subtitle item. Change the Text property to Date: 07/10/07, and change the font to Arial 10.

8   Click Field Names and change to font to Arial Bold 12 Underline.

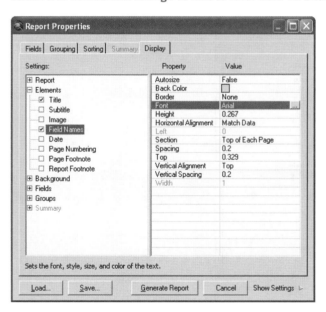

9   Expand the Fields list and change the field names to **Airport Name**, **Airport ID**, and **State**.

10 Change all fonts to Arial 9, and change the width of the Name field to 3.

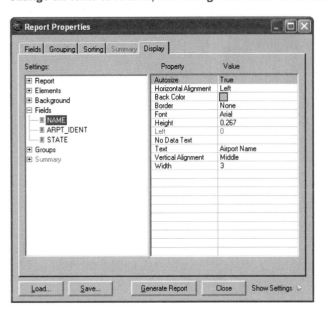

11 Expand the Groups list and click STATE. Change the Back Color to gray, and the font to Arial Bold
    14 Black Underline, and the width to 6.

EXERCISES

12 **Click Generate Report.**

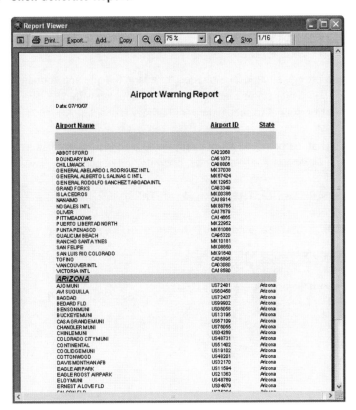

The Airport Warning Report is generated and shown in the Report Viewer. The report lists all the airports in the United States, Canada, and Mexico within 800 miles of the targeted threat site. This report can be used to notify personnel at these locations to step up surveillance to detect suspicious activity at their airfield, such as transporting and storing a commercial rental trailer near the airpark, or installing a new spray system onto a newly acquired airplane.

This report can be printed, exported to a PDF, RTF or TXT file and distributed either electronically or on paper to personnel at the selected facilities. It can also be added to a map layout of the locations of the facilities and distributed accordingly.

13 **When you are satisfied with the report format, click Export from the Report Viewer tool bar.**

14 **Export your report as FL_Airport_Warning_Rpt.pdf, with FL being your initials, in \ESRIPress \GISTHS\MYGISTHS_Work.**

15 **Close the report, and save this report format as FL_Airport_Warning_Rpt.rdf, with FL being your initials, in \ESRIPress\GISTHS\MYGISTHS_Work.**

YOUR TURN

Select by attributes only the airports in California. Generate a report of only these selected airports by checking the Use Selected Set in the Fields tab of the Report Properties window.

Exercise 4.3
Build warning map layout

Maps produced in the *Prevent* Mission Area provide situational awareness of a targeted site and surrounding area. Preevent planning is most effective when visual status maps of the MEDS, coupled with data layers relevant to a particular threat situation, alert personnel to circumstances that may warrant closer surveillance.

The warning maps prepared in this exercise can be produced as interactive map displays, image files for placement in presentation documents such as reports and slide shows, or printed on paper for display.

1   Add USA_States.shp from \ESRIPress\GISTHS\GISTHS_C4 to FL_GISTHS_C4E2.mxd.

2   Select by Location the USA_States that intersect with the FL_Airports_800m data layer.

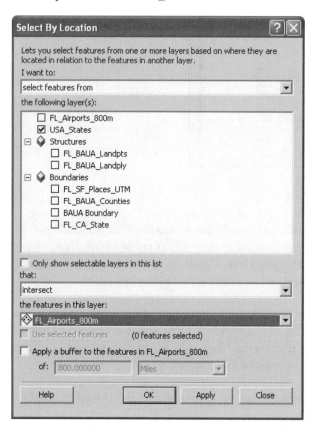

3   Export the selected states as **FL_States_ Airports_800m.shp**, with FL being your initials, save it in the **\ESRIPress\GISTHS\MYGISTHS_Work** folder and add it to the map.

4   Remove USA_States from the Table of Contents.

5   Zoom to the extent of FL_States_Airports_800m and label the states using STATE_NAME as the label field.

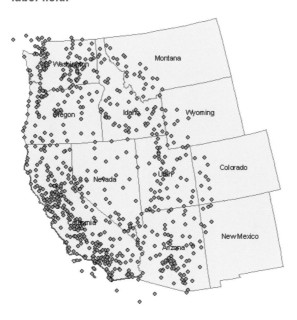

6   For another reference to this map, add USA_Interstates.shp from **\ESRIPress\GISTHS\GISTHS_C4**.

7   Use the Clip tool in ArcToolbox to clip this interstates layer to the extent of the FL_States_Airports_800m layer. Save this new layer as **FL_Interstates_Airports_800m.shp** in **\ESRIPress \GISTHS\MYGISTHS_Work**.

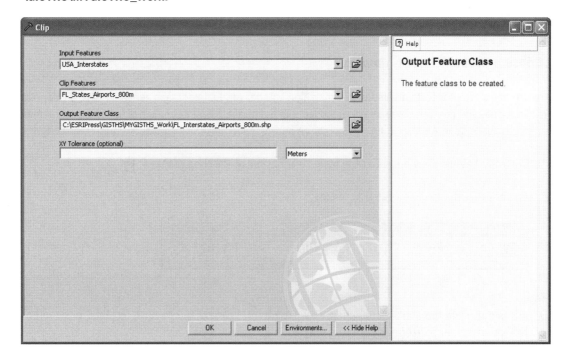

EXERCISES

8   Close the Clip box when completed, and remove USA_Interstates from the Table of Contents.

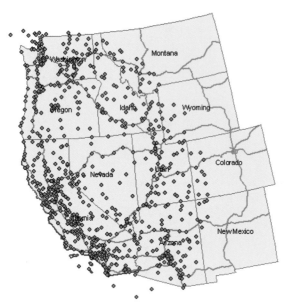

All of the airports within 800 miles of the targeted site are shown on the map. Adding the states and interstate highways to the map provide geographic reference to the map display to enable homeland security personnel to identify more specifically where these airport facilities are located.

9   Save the ArcMap map document as **FL_GISTHS_C4E3a.mxd,** with FL being your initials, to the **\ESRIPress\GISTHS\MYGISTHS_Work** folder.

Choose a prebuilt layout template

ArcMap has a layout view that sets up your map elements for production. Maps generally have a title, map, legend, scale, and possibly some other components to define the map information. There are several map templates available in ArcMap to help you organize and produce maps using standard map formatting.

1   Click the Layout View button at the lower left corner of the map display.

2   Be sure that the Layout toolbar is visible.

3   If it is not, from the Main Menu, click View, Toolbars, Layout, and be sure to place this toolbar at the top of the map display where it is fully visible.

4   From the Layout Toolbar, click the Change Layout button.

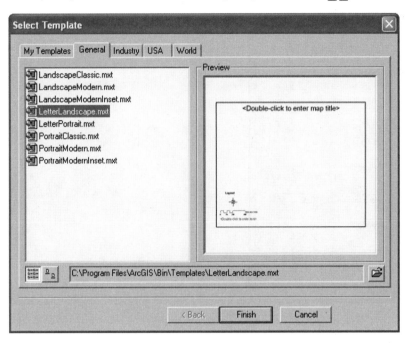

The Select Template window opens.

5   Click the General tab, and select the LetterLandscape.mxd template, if it is not already selected.

6   Click Finish. Use the Zoom and Pan tools to better position your map in the layout.

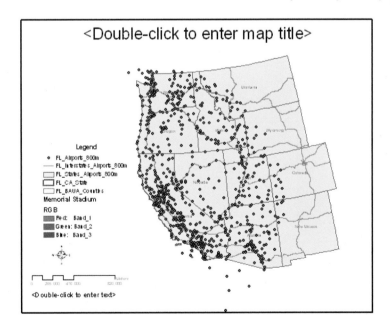

Your map is placed in this template where you can customize the title and legend, and add text to further describe your map.

7   Double-click to enter the map title Airport Warning Map.

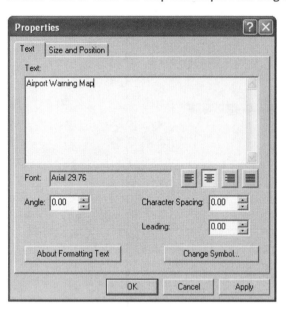

8   Click OK.

9   Double-click the scale bar and change the units to miles.

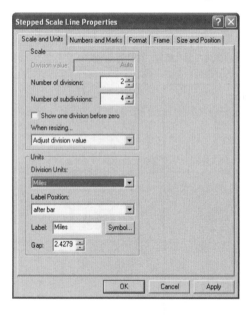

10   Click OK.

11   Drag the right end of the scale bar to stretch it out to 800 miles. Use the Zoom tools on the Layout toolbar to zoom in closer to see the scale bar more clearly.

12   Double-click the Legend box to open the Legend Properties window.

13 Include only the FL_Airports_800m and FL_Interstates_ Airports_800m data layers in the Legend Items box.

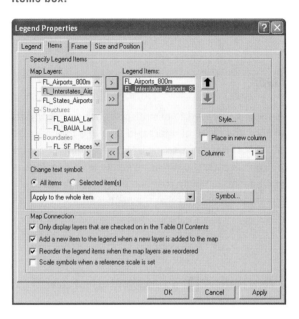

14 Click OK.

15 Double-click the text box below the scale bar and add your name and date.

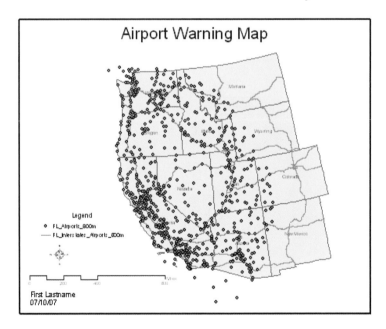

This map was generated from a prebuilt layout that quickly placed all of the relevant map elements into a suitable layout for display. Accompanied by the warning report created earlier, it can be distributed to homeland security personnel to alert identified airports of the possibility of an attack originating from their facility. Steps can then be taken to reinforce surveillance of those facilities and surrounding areas.

16 Save FL_Airports_800m, FL_States_Airports_800m, and FL_Interstates_ Airports_800m as layer files in **\ESRIPress\GISTHS\MYGISTHS_Work** for use in the next exercise.

17 Save and close your map document.

Create a custom map layout for multiple maps

Sometimes a map template doesn't contain all of the elements needed to best display your map. For instance, if you want to include a zoomed map in your layout to show the detail of a targeted area in a region, you can create additional data frames and include them in one layout. You can also include a report and/or a graph in the layout as a map element. At the same time, you can save a custom map template to use again for a similar set of data layers. In this exercise, you will create a map template that can be used to produce a series of maps, one for each state where there are airports to be notified that a potential attack may originate from their location.

1 Open a new map document.

2 If not already in layout view, click the Layout View button ☐ to switch to layout view.

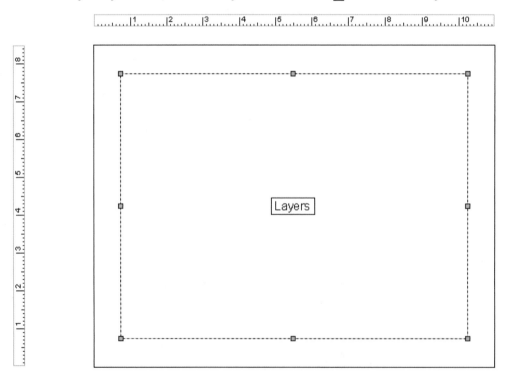

A layout screen appears indicating the placement of the empty Layers data frame. Note the guide rulers along the edge of the layout. They enable the alignment of objects to ensure a balanced layout.

3   **Click on the half-inch rule mark along the horizontal guide ruler at the top left of the layout window.**

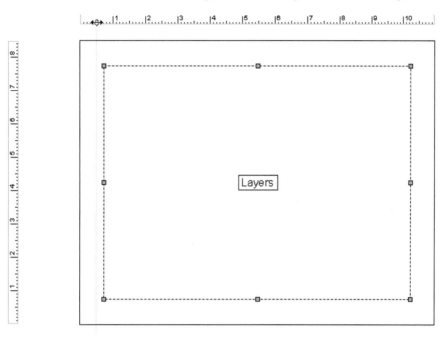

A light blue vertical grid line appears on the layout.

4   **Click the 10½-inch rule mark at the top right; the half-inch mark at the bottom left; and the 8-inch mark at the top left.**

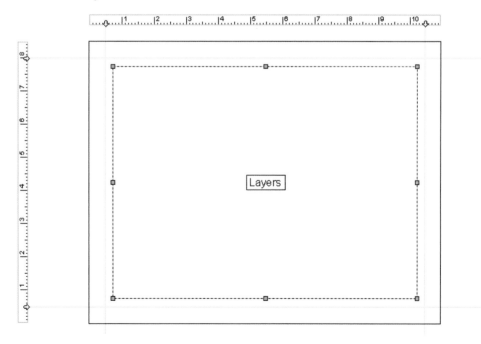

You have created a border around the map layout to ensure that no objects that are placed in the layout extend beyond the marked border.

5   Click in the center of the data frame box and drag it up to the new grid lines at the upper right corner of the map layout. Notice that it snaps into place.

6   Drag the lower left corner of the box to resize the box to one-quarter of its original size.

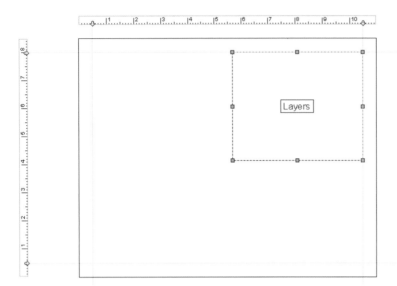

7   Add the FL_Airports_800m, FL_Interstates_Airports_800m, and FL_States_800m data layers to the data frame.

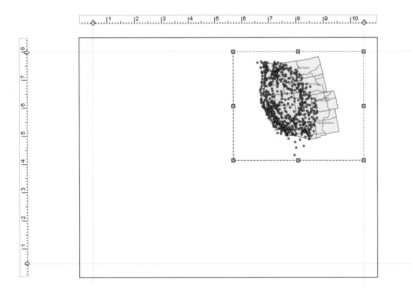

This map layout now shows the data layers in a small window. There is room on the layout to include other data frames that will add location context to the map.

8   Rename the data frame in the Table of Contents from Layers to Airports.

9   Save the ArcMap map document as **FL_GISTHS_C4E3b.mxd**, with FL being your initials, to the **\ESRIPress\GISTHS\MYGISTHS_Work** folder.

Add multiple data frames to a map layout

1   From the Main Menu, click Insert and select Data Frame.

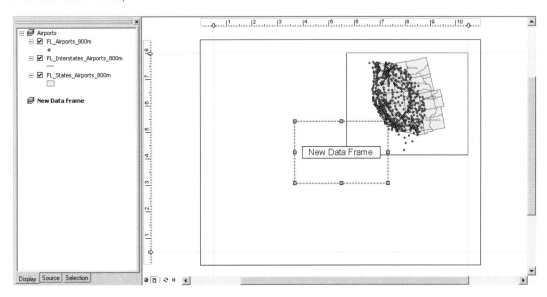

If you are in the data view, the screen will go blank. If you are in layout view, a new data frame box appears on the map layout. The new data frame has a blue highlighted box around it, and is bolded in the table of contents to show that it is now the active data frame. You can toggle between active data frames by either clicking on each one in the layout display, or by right-clicking on the data frame in the table of contents and selecting activate.

2   Add the **FL_BAUA_MEDS_Orthoimagery** data layer to this new data frame.

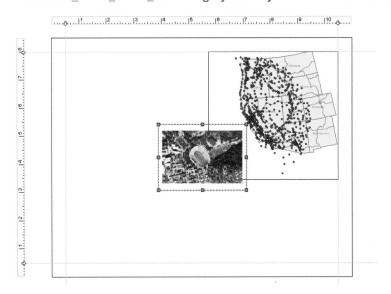

3   Rename the new data frame **Memorial Stadium**.

4   Return to the data view and right-click on the Airports data frame in the Table of Contents and select Activate to activate this data frame.

5   Use the Select Features Tool 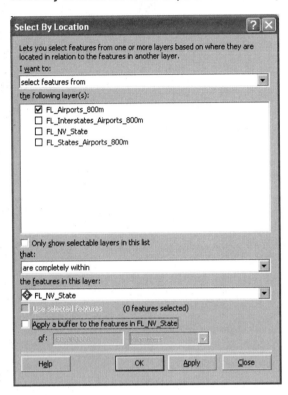 to select the state of Nevada on the map.

6   Export this selected feature from FL_States_Airports_800m as FL_NV_State.shp with FL being your initials, to \ESRIPress\GISTHS\MYGISTHS_Work, and add it to the map.

7   Select by location all of the airports that are completely within the state of Nevada.

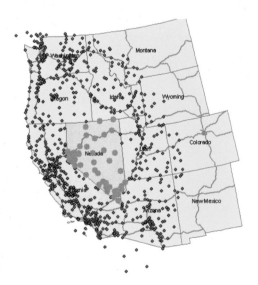

8   Export these selected features as **FL_Airports_800m_NV.shp**. to **\ESRIPress\GISTHS\MYGISTHS_Work**, and add it to the map.

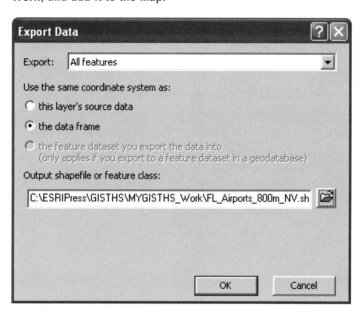

9   Clear all selected features.

10   Insert a new data frame, and rename it **Nevada**.

11   Drag the FL_NV_State, FL_Airports_800m_NV and FL_Interstates_Airports_800m data layers from the Airports data frame into the Nevada data frame.

12   Label the features of all of the data layers in the Nevada data frame using unique fonts and colors so they are distinguishable from each other.

13   Change the symbol marker for the airports to Airfield Size 10.

14 Return to the layout view.

15 Arrange the data frames in the layout as shown below. Add additional grid lines to act as guides on the layout to align the data frames. Use the Pan tool 🖑 inside the Airports data frame to shift the map to the right to provide room to place the Memorial Stadium data frame inside of it.

Add a callout box to a map layout

1 To indicate the location of the stadium on the small scale Airport map, click the drop-down box next to the New Text button **A** on the Drawing toolbar to drag a callout box from the BAUA to the orthoimage of the stadium.

2 Double-click the callout text box and enter **Memorial Stadium**.

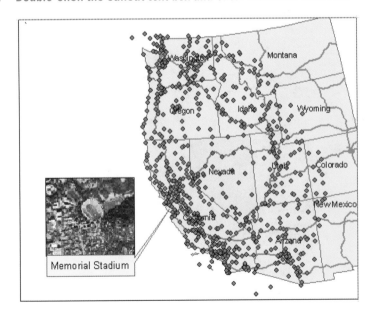

Insert map layout elements

Now that the data frames of this map document have been placed in the map layout, you can add other map elements as needed to complete the layout design. Other elements include a legend, title, scale, directional arrow, additional text, and tables and graphs. Each of these elements can be added individually and customized to the particular specifications of the map document.

1   From the Main Menu, click Insert, and select Title. A title text box appears filled with the filename.

2   Right-click the title text box, and select properties.

3   Enter the new title as **Airport Warning Map** and click Change Symbol to select font Arial Bold 24.

4   Click OK.

5   From the Main Menu, click Insert, and select Neatline. Choose Place Inside Margins. Select a double line border and None for background.

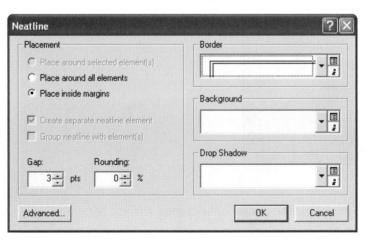

6   Click OK.

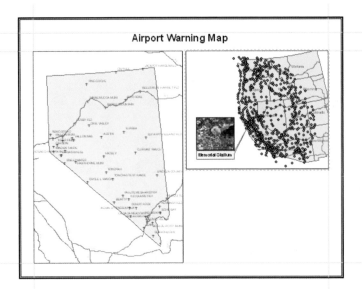

A neatline is placed around all of the map elements in the map layout.

7   Click the Nevada data frame box to make it the active data frame.

8   From the Main Menu, click Insert and select Legend.

9   Remove FL_NV_State from the Legend Items list.

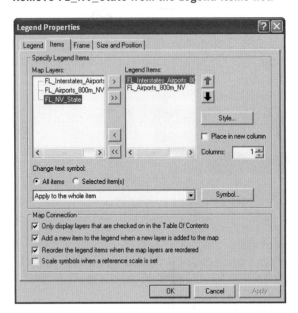

10   Click Next four times and then click Finish to complete the Legend Wizard.

11 Drag the Legend to the lower right of the Nevada map, and snap it in place to the grid line.

12 From the Main Menu, click Insert and select North Arrow.

13 Select an appropriate north arrow and place it to the right of the legend.

14 Ensuring that the Nevada data frame is active, from the Main Menu, click Insert and select Scale Bar.

15 Select an appropriate scale bar.

16 Click Properties and change the units to Miles.

17 Click OK twice.

18 Position the scale bar within the Nevada data frame and resize it to 120 miles.

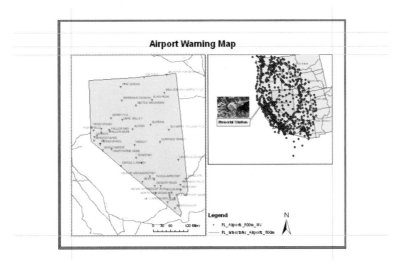

The scale bar that is inserted will be for only the active data frame. Since there are multiple data frames in this map layout, but sure that you are using a scale bar for the appropriate data frame.

19   From the Main Menu, click Insert, and select Text.

20   Enter your name and date, and change the font to Arial 12. If you cannot see the text box clearly, zoom into the layout or open the properties window to edit the text.

21   Drag the text box to the lower right corner of the map layout and snap it in place to the guide lines.

Your airport warning map is now complete. It can be used to alert homeland security personnel in Nevada that there are airports around the state within the range of a potential point of origin for an attack on Memorial Stadium.

22   Save the ArcMap map document as **FL_GISTHS_C4E3c.mxd**, with FL being your initials, in **\ESRIPress\GISTHS\MYGISTHS_Work**.

Save a custom map template to My Templates

This map layout can be saved and used again to create maps for other states that have airports within 800 miles of the potential attack site. This enables all of the state maps to be alike and reduces the time needed to produce the map series.

1   From the Main Menu, select File, Save As.

2   Navigate to your operating system user profile folder at **/Documents and Settings/[user profile] /Application Data/ESRI/ArcMap/Templates**.

3   Select ArcMap Template (*.mxt) as the Save as type.

4   Save this template as **FL_Airport_Warning_Map.mxt**, with FL being your initials.

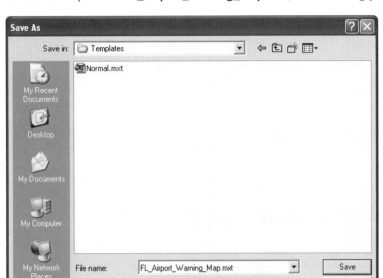

Because you saved this template in your operating system user profile folder, it will appear under the My Templates tab in ArcMap. You can then select this custom map template to use in the future.

5   Close ArcMap.

Use a custom map template

Now that you have created a custom map template, you can use it to generate a series of maps. It may not be necessary for the homeland security personnel in Oregon to have a map that shows the location of airports in Nevada. Using this custom template, you can quickly create another map using the same map elements, but show data of Oregon rather than Nevada.

1   Open ArcMap to a new map document.

2   If not already in layout view, click the Layout View button to switch to layout view.

3   Click the Change Layout button 🖼 on the Layout toolbar.

4   Click the My Templates tab, and select the FL_Airport_Warning_Map.mxt template in the window.

5   Click Next.

6   Click Finish to load the map template.

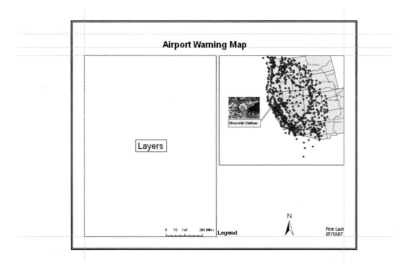

All of the map elements load from the template, but the data frame containing the airports for the individual state is empty.

7   Activate the Airports data frame in the Table of Contents.

8   Select by attribute the state of Oregon from the FL_States_Airports_800m layer and export it as **FL_OR_State**, with FL being your initials, in **\ESRIPress\GISTHS\MYGISTHS_Work**.

9   Click Yes to add it to the map as a layer, and clear selected features.

10   Select by attributes all of the airports in Oregon from the FL_Airports_800m layer, and export it as **FL_Airports_800m_OR**, with FL being your initials, in **\ESRIPress\GISTHS\MYGISTHS_Work**.

11 Click Yes to add it to the map as a layer, and clear selected features.

12 Drag these new layers into the Layers data frame.

13 Drag the FL_Interstates_Airports_800m into the Layers data frame.

14 Change the symbol marker for the airports to Airfield Size 10.

15 Label the Airport data layer.

16 Click the Layers data frame in the map layout to make it active.

17 Delete the scale bar and insert a new one to ensure that the scale bar is associated with the correct data frame.

18 Remove FL_OR_State from the legend.

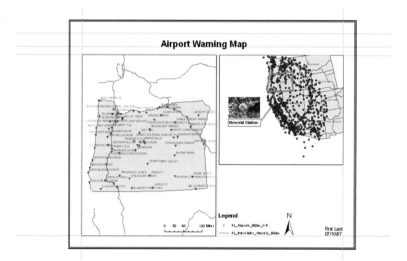

This map can now be sent to homeland security personnel in Oregon to alert them that there are airports around the state within the range of a potential point of origin for an attack on Memorial Stadium. Using a map template to standardize map production makes turning out multiple maps of similar data fast and efficient. Accompanying this map with a detailed report of contact information for each airport enables security personnel to quickly notify staff to step up surveillance and possibly prevent an attack.

19 Save the map document as **FL_GISTHS_C4E3d**, with FL being your initials, in **\ESRIPress\GISTHS \MYGISTHS_Work**.

YOUR TURN

Use the custom template to create an airport warning map for airports in another state.

Add a report to the map layout

It is also possible to include a tabular report directly in a map layout itself, rather than accompanying the printed map as a separate document.

1   In order to make room in the layout for the report, remove all data frames except the Layers data frame.

2   Generate a report of the airports in Oregon.

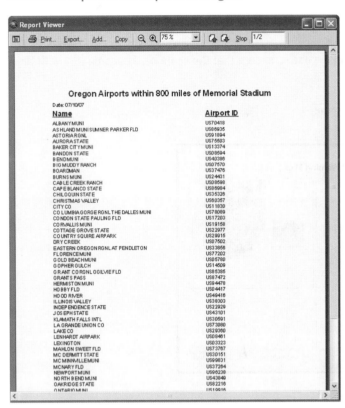

3   Click the Add button in the Report Generator to add this report to the map layout.

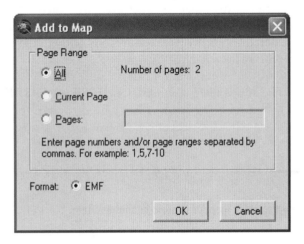

4 Click OK and close the Report Viewer.

5 Save the report as **FL_Airport_Warning_OR**, with FL being your initials, to **\ESRIPress\GISTHS \MYGISTHS_Work**.

6 Close the Report Properties window.

7 Resize and reposition the report pages and other map elements as needed to fit them all on the map layout.

This map now contains all the map elements needed to identify the airports in Oregon that could be potential points of origin for an attack on Memorial Stadium, as well as a detailed report showing the airport names and identification codes.

8 Save the map document as **FL_GISTHS_C4E3e**, with FL being your initials, in **\ESRIPress\GISTHS \MYGISTHS_Work**.

Exercise 4.4
GIS outputs

These maps and reports can be disseminated to homeland security personnel in a number of effective ways. The most traditional output is the printed map. ArcMap enables a map to be formatted for a wide variety of print options, including report-sized layout, to large format poster size plotter output.

Print report-sized layout

1   From the Main Menu, click File and select Print.

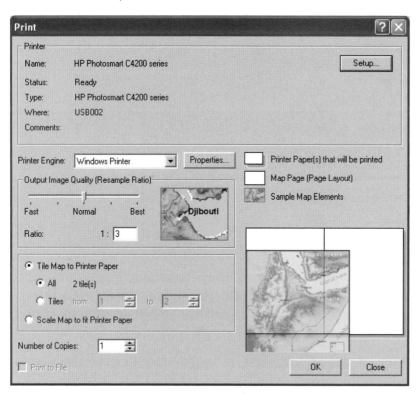

Your Windows printer is selected as the printer engine. The graphic in the print window shows that the layout is set to the portrait orientation, while the map is best displayed in landscape mode. Your printer page may show a slightly different map display, depending on your printer type.

2   Click Setup and set the paper orientation to Landscape. Click OK.

3   Click Scale Map to fit Printer Paper.

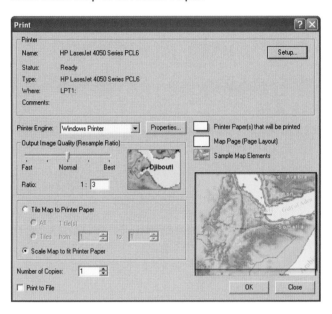

The graphic now shows the layout in landscape mode.

4   Click OK to send your layout to the printer.

Print large poster-sized plot

If you have a plotter for printing large poster sized maps, you can set the print options in ArcMap to scale a layout to these dimensions.

1   From the Main Menu, click Page and Print Setup.

2   To plot your map at a standard E size, select ARCH E as the page size.

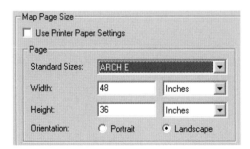   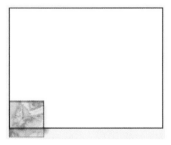

The sample map now sits at the lower left corner of the layout graphic.

3   Click Scale Map Elements proportionately to changes in Page Size.

4   Click OK.

5   If you have a large output plotter to print your map, click Print from the File menu and send your map to your plotter.

6   Reset the map page size to Letter.

Export layouts

ArcMap allows you to export your map layout in a variety of formats. An exported layout can be included in a Microsoft Word or PowerPoint document, or can be uploaded to a Web page for Internet viewing.

1   From the Main Menu, click File, Export Map.

2   Select TIFF (*.tif) as the type, and save the output as **FL_GISTHS_C4E3e.tif**, with FL being your initials, in **\ESRIPress\GISTHS\MYGISTHS_Work**.

3   Click Save.

Your map layout is now saved as an image file that can be easily imported into a working document or Web page for viewing.

Other outputs

Other output capabilities in ArcMap include using the ArcGIS Publisher extension, which enables the creation of published map files from any ArcMap document that can be viewed and interacted with simultaneously over local or Internet-based networks.

Additional output capabilities extend to ArcGIS Server, which is a comprehensive server-based GIS that enables the delivery of GIS functionality and outputs throughout an enterprise system, such as a regional integrated homeland security GIS operation.

Summary

In this chapter, you learned the basic ArcMap skills used to prepare and produce reports and map layouts. Map layouts are composed of a suite of elements that define the map information presented. Prepared and custom map templates help create effective warning reports and map layouts that can be printed on paper or exported in a variety of digital formats for electronic distribution. Within the *Prevent* Mission Area of homeland security pre-event planning, this set of skills is essential to disseminating critical information that may actually prevent an attack.

Notes

1. *National Strategy for Homeland Security*, p. 2

2. *Target Capabilities List: Version 1.1*, U.S. Department of Homeland Security, Office of State and Local Government Coordination and Preparedness, May 23, 2005, p. 30.

3. *Universal Task List: Version 2.1*, Department of Homeland Security, Office of State and Local Government Coordination and Preparedness, May 23, 2005, p. 27

4. Ibid., p. 25

5. *National Planning Scenarios, Version 20.1*, April 2005, p. 5-1

6. Ibid., p. 5-5.

7. Ibid., p.5-4.

Chapter 5 (5.1-5.3):

Analyzing data for homeland security planning and operations—*Protect*

DHS National Planning Scenario 12: *Explosive attack— bombing using improvised explosive devices*

5.1   Prepare protect scenario map 1

5.2   Locate critical infrastructure
Select by location (completely within)
Summarize data in a table

5.3   Protect critical infrastructure
Create protective security buffer zone around critical
    infrastructure
Secure ingress and egress routes around critical
    infrastructure
Site surveillance locations around critical infrastructure
Generate hillshade surface
Create new shapefile of security points in ArcCatalog
Add new shapefile to map document
Generate viewshed
Create line of sight
Create line-of-sight profile graph
View 3D line of sight

MISSION AREA: *PREVENT*
Detection in preevent planning
stages.

MISSION AREA: *PROTECT*
DHS Target Capability: *Critical
infrastructure protection program*
DHS Task: *Pro.B1 implement
protection measures*

MISSION AREA: *RESPOND*
Alerts, activation, notification,
traffic and access control,
protection of special populations,
resource support, requests for
assistance, and public information.

MISSION AREA: *RECOVER*
Decontamination, removal, and
disposal of debris and remains;
repair and restoration of main
venue and transportation center.

Define *protect*, to include safeguarding critical infrastructure and
the safety of people impacted by emergency events
Locate critical infrastructure
Use 2D and 3D spatial tools to site optimal surveillance positions
around critical infrastructure
Map demographics to locate and identify population potentially
impacted by an emergency event
Locate assembly points for evacuation outside of an impacted area

Chapter 5

Analyzing data for homeland security planning and operations: *Protect*

This chapter focuses on using GIS tools and analysis to support the *Protect* Mission Area as identified in the DHS National Preparedness Goal. Protecting the nation's people and places from the effects of man-made and natural disasters entails identifying the most likely targets for attack or disaster as well as who and what may be impacted by such events.

As early as 1996, President Bill Clinton signed Executive Order 13010, establishing the President's Commission on Critical Infrastructure Protection.[1] At that time, critical infrastructure was defined as those "... national infrastructures ... so vital that their incapacity or destruction would have a debilitating impact on the defense or economic security of the United States."[2] In May 1998,

Presidential Decision Directive No. 63 (PDD-63) further defined critical infrastructure to include "those physical and cyber-based systems essential to the minimum operations of the economy and government."[3]

Since then, additional executive orders and presidential directives have further clarified what critical infrastructure is and in what sectors of the economy and government they reside. The traditional notion of limiting critical infrastructure to physical structures, such as transportation networks, water systems, and energy utilities gave way in the USA PATRIOT Act of 2001 to include information and telecommunications, banking and finance, and public health and safety. In 2003, the Homeland Security Presidential Directive No. 7 (HSPD-7), *Critical Infrastructure Identification, Prioritization, and Protection*, went so far as to analyze critical infrastructure not by physical construct, but by economic sector to include "agriculture, food, water, public health, emergency services, government, defense industrial base, information and telecommunications, energy, transportation, banking and finance, chemical industry, and postal and shipping."[4]

Defining what makes infrastructure critical was expanded to include not only disruptions to the economy and defense but also the "essentials for accomplishing missions affecting life and property."[5] Put simply, critical infrastructures are those people, things, or systems that must be intact and operational in order to make daily living and working possible.[6]

The National Asset Database (NADB), as discussed in chapter 3, is designed to be a comprehensive listing of all national assets that may be targeted for attack by terrorists. As of January 2006, the classified database contained 77,069 sites, including 17,327 office buildings, malls, shopping centers, and other commercial properties; 12,019 government facilities; 8,402 public health buildings; 7,889 power plants; and 2,963 sites with chemical or hazardous materials.[7]

It is obvious, however, that not all critical infrastructures are equally critical. The first generation NADB has been highly criticized for its inclusion of many noncritical assets and has been deemed an inaccurate representation of the nation's critical infrastructure/key resources (CI/KR).[8] The second generation of the NADB promises to include a comprehensive risk assessment to distinguish critical from noncritical resources, so that it can support the management and funding allocation decision-making process envisioned by DHS.

Vulnerability or risk assessment determines if and when a key resource is truly critical to the survivability and sustainability of a community or the nation as a whole. Certain key resources may only be critical at certain times, such as a stadium that remains empty most of the time but becomes a target when it is occupied by tens of thousands of spectators attending a major event with worldwide exposure. Other key resources, such as the Golden Gate Bridge, may be under threat at all times, not so much because its destruction would impact the transportation system in the San Francisco Bay Area, but more significantly because of its perception as a national icon.

Vulnerability assessment is an uncertain science, at best. There are a number of assessment tools and methodologies that assist security planners in determining what key resources in their communities are vital and critical at all, or particular times. The CIP Process methodology outlined in the DHS United States Fire Administration Job Aid document is a "decision sequence that assists leaders in ultimately determining exactly what really needs protection."[9] The process consists of the following steps:

- Identifying critical infrastructures
- Determining the threat
- Analyzing the vulnerabilities
- Assessing risk
- Applying countermeasures

The Job Aid guide includes a CIP Process Question Navigator and Decision Matrix to help planners work through the evaluation.

An additional assessment resource includes the CARVER2 risk-assessment software that ranks critical infrastructure in order of importance by providing for cross-sector scoring and ranking. It is freely available to government agencies as a PC-based version or Web-based application that integrates real-time mapping via GPS and USGS geospatial data.[10]

In addition to protecting the nation's infrastructure from attack, homeland security planners are tasked with developing protection plans for people potentially affected by a catastrophic event. Populations with special needs are of particular importance as they are often unable to evacuate without assistance. GIS tools and analysis are very effective in this task. Census data can be spatially analyzed to identify where certain populations, such as young children or elderly people, reside within close proximity to an event site.

Chapter 5: Data dictionary

Layer	Type	Layer Description	Attribute	Description
Petrol_Tanks.shp	Shapefile	Petrol tank location points		
Petrol_Tanks.lyr	Layer File	Petrol tank location points layer file		
PLUME_COBALT_HYDROCARBONYL.shp	Shapefile	Plume extent polygon		
PLUME_COBALT_HYDROCARBONYL.lyr	Layer File	Plume extent polygon layer file		
tgr06000sf1blk.dbf	dBase Table	2000 Census block polygons	STFID	Census block ID code
			POP2000	2000 Census block population
			Age_65_UP	2000 Census block population age 65 and up
tgr06013blk00.shp	Shapefile	2000 Census block demographics	STFID	Census block ID code

Exercise 5.1
Prepare protect scenario map 1

DHS National Planning Scenario 12:
> Explosive attack—bombing using improvised explosive devices

DHS Target Capability:
> Critical infrastructure protection (CIP)

DHS Task:
> Pro.B1 Implement protection measures

In this scenario, a threat is made targeting a stadium in an urban area during an upcoming event. Due to the nature of this threat, the stadium site is regarded as the primary critical infrastructure and a number of key resources in and around the stadium site are considered secondary critical infrastructure.

DHS National Planning Scenario 12: Explosive attack—bombing using improvised explosive devices, is described as follows:

> In this scenario, agents of the Universal Adversary (UA) use improvised explosive devices (IEDs) to detonate bombs at a sports arena and create a large vehicle bomb (LVB). They also use suicide bombers in an underground public transportation concourse and detonate another vehicle bomb in a parking facility near the entertainment complex. An additional series of devices is detonated in the lobby of the nearest hospital emergency room (ER).[11]

The scenario profile includes a synopsis of how the Protect Mission Area is to be activated in response to this threat:

> The planning and execution of the event would require a significant level of relatively unsophisticated coordination. As such, the potential for detection in the pre-event planning stages exists. The completion of a targeting package would necessitate obtaining or creating diagrams of the venue, the transportation platform, the hospital ER, and the environments around these sites. Surveillance of the target location would be conducted, with photographs and video documentation performed.[12]

The DHS FY 2007 Infrastructure Protection Program (IPP) includes the Buffer Zone Protection Program Award. The purpose of the grant is to provide a total of $48.5 million in funding to build security and risk-management capabilities at the state and local level to secure predesignated Tier I and Tier II critical infrastructure sites, including chemical facilities, financial institutions, nuclear and electric power plants, dams, stadiums, and other high-risk/high-consequence facilities.[13]

The use of GIS tools in this scenario is vital. HSPD 7 specifically states that the "Secretary will collaborate with other appropriate Federal departments and agencies to develop a program, consistent with applicable law, to geospatially map, image, analyze, and sort critical infrastructure and key resources ..."[14] By mapping a buffer zone around the critical infrastructure site, homeland security planners can meet the DHS task to implement protective measures by: 1) locating key critical infrastructure within an extended buffer zone beyond the specific site; 2) securing the perimeter of a protective buffer zone around a targeted site by stepping up surveillance posts at ingress and egress points; and 3) siting the most optimal observation positions in and around an event site.

Add MEDS data to scenario map

1 Open a new ArcMap document.

2 Add the FL_BAUA_MEDS_Boundaries, Orthoimagery, and Geographic Names group layers from the \ESRIPress\GISTHS\MYGISTHS_Work folder.

3 Zoom to the extent of the orthoimagery layer.

4 Turn off the labels of all data layers, if they appear.

5 Set selectable layers to only the FL_BAUA_GNIS data layer.

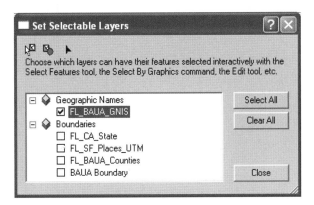

6 Use the Select Tool  to select the marker for Memorial Stadium.

Create buffer around extended area beyond event site

1   Open ArcToolbox.

2   Expand Analysis Tools and Proximity in the ArcToolbox window.

3   Double click Buffer.

4   Select FL_BAUA_GNIS as the Input features; Output feature class as **\ESRIPress\GISTHS** **\MYGISTHS_Work\FL_BAUA_GNIS_Buffer.shp**; and set the Linear unit to 3 miles.[15]

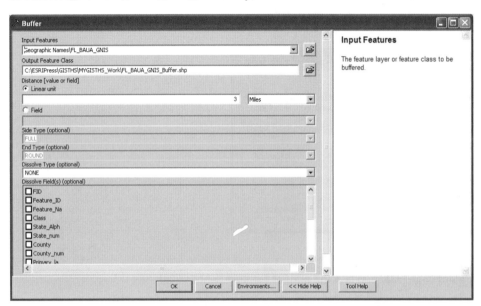

5   Click OK.

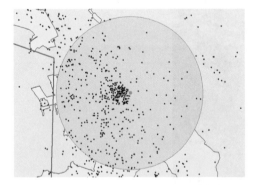

6   Close the Buffer window when processing is complete.

7   Zoom to the new FL_BAUA_GNIS_Buffer data layer, and close ArcToolbox.

The buffer extends three miles in all directions around the Memorial Stadium site.

8   Save the ArcMap document as **FL_GISTHS_C5E1.mxd**, with FL being your initials, in **\ESRIPress** **\GISTHS\MYGISTHS_Work**.

Exercise 5.2
Locate critical infrastructure

Now that you have prepared the scenario map to show the extent of the designated buffer zone around the targeted site, you can now locate those key resources within this zone that may be potential secondary targets. These sites will also require additional security measures to step up surveillance to monitor activities in and around them to prevent terrorists from staging secondary incidents that may threaten the safety of the people in the area.

Select by location (completely within)

1   From the Main menu, select Selection, and click Select by Location.

2   Check the box next to FL_BAUA_GNIS to select features from.

3   Select are completely within and FL_BAUA_GNIS_Buffer.

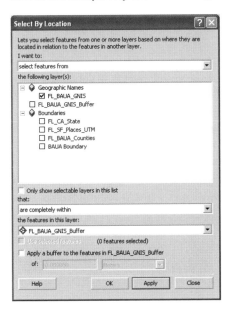

4   Click OK.

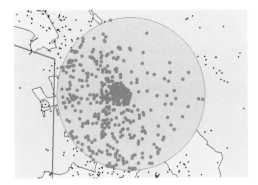

All of the GNIS points within the buffer are selected and highlighted.

EXERCISES

5   Open the FL_BAUA_GNIS attribute table and view the selected records.

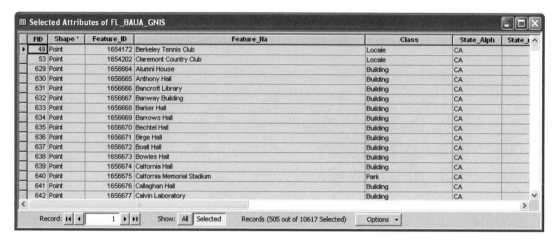

There are 505 features selected within this buffer zone.

Summarize data in a table

It is difficult to identify what features are included in this selected set by simply viewing the table. Summarizing the data in this table by Class tallies the number of features in each class. Homeland security planners can then view the list to identify which features may be included in a vulnerability assessment to determine if they are critical infrastructure given the nature of the proposed security threat.

1   Right-click the Class field heading and select Summarize.

2   Specify the output table as **FL_GNIS_Buffer_Features.dbf**, with FL being your initials, in **\ESRIPress\GISTHS\MYGISTHS_Work.**

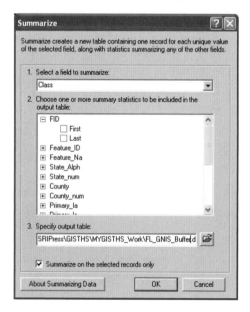

EXERCISES

3 Be sure that the box beside Summarize on the selected records only option is checked.

4 Click OK.

5 Click Yes to add the result table in the map.

6 Close the Attributes of FL_BAUA_GNIS table.

7 Click the Source tab at the bottom of the Table of Contents and open the FL_GNIS_Buffer_ Features table.

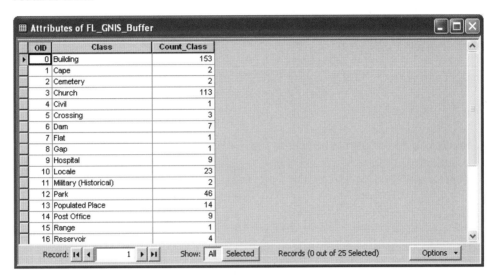

The features are listed alphabetically showing the number of features within each class. Certain feature classes are considered critical infrastructure given the conditions of this particular threat scenario. Hospitals and locales (where, in this case, train stations are listed) are particularly important in this scenario, as secondary IEDs are threatened to be detonated at these critical locations after the initial explosion at the stadium. Stepping up pre-event security and surveillance at these locations may make it increasingly difficult for adversaries to plan and stage these secondary incidents.

Additional key resources within this buffer zone include buildings, especially university laboratories that may contain hazardous materials; reservoirs and dams; tunnels; and other features that may be impacted by such an incident.

8 Close the Attributes of FL_GNIS_Buffer_Features table.

9 Save the ArcMap document as **FL_GISTHS_C5E2.mxd**, with FL being your initials, in **\ESRIPress \GISTHS\MYGISTHS_Work**.

EXERCISES

Exercise 5.3
Protect critical infrastructure

GIS tools are vital to design and implement protective measures that secure critical infrastructure. Buffer zones around key locations demarcate optimal surveillance locations at ingress and egress points; and viewshed and line-of-sight analysis assist in siting the best possible observation positions in areas of terrain relief in and around an event site.

Create protective security buffer zone around critical infrastructure

1   Open the Attributes of FL_BAUA_GNIS table.

2   Click the Options button.

3   Click Select by Attributes.

4   Select Method: Select from current selection.

5   Enter the expression "Class" = 'Hospital'.

6   Click Apply.

7   Click Selected to view only the selected records.

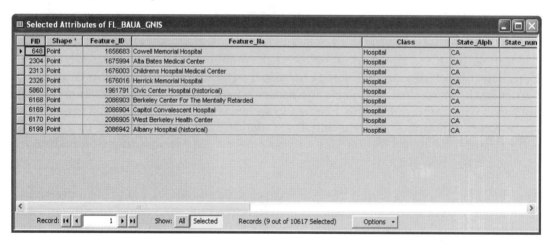

FID	Shape	Feature_ID	Feature_Na	Class	State_Alph	State_num
648	Point	1656683	Cowell Memorial Hospital	Hospital	CA	
2304	Point	1675994	Alta Bates Medical Center	Hospital	CA	
2313	Point	1676003	Childrens Hospital Medical Center	Hospital	CA	
2326	Point	1676016	Herrick Memorial Hospital	Hospital	CA	
5860	Point	1961791	Civic Center Hospital (historical)	Hospital	CA	
6168	Point	2086903	Berkeley Center For The Mentally Retarded	Hospital	CA	
6169	Point	2086904	Capitol Convalescent Hospital	Hospital	CA	
6170	Point	2086905	West Berkeley Health Center	Hospital	CA	
6199	Point	2086942	Albany Hospital (historical)	Hospital	CA	

Record: 1   Show: All Selected   Records (9 out of 10617 Selected)   Options ▼

There are nine hospitals within this buffer zone that could be potential staging sites for the secondary incidents planned in this scenario. This list can be distributed to security personnel as they prepare plans to protect these key resources.

8   Close the attribute table to view the selected hospital locations on the map.

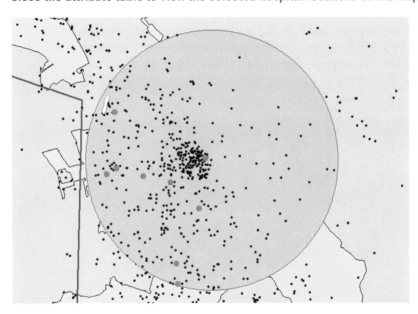

9   Generate a 500-foot buffer around these hospital points to create a security perimeter surrounding each site. Specify the output table as **FL_BAUA_Hospital_Buffer500.shp**, with FL being your initials, to **\ESRIPress\GISTHS\MYGISTHS_Work**.

10 Clear selected features.

11 Drag this new buffer layer to the top of the Table of Contents.

12 Increase the transparency of this buffer layer to 50%.

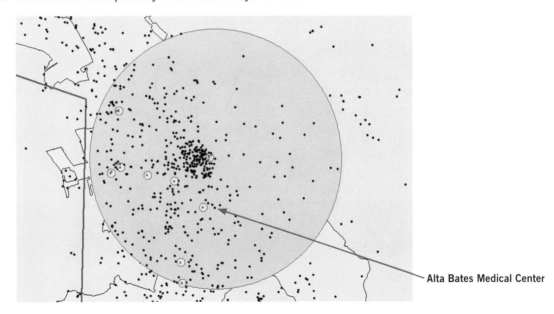

Alta Bates Medical Center

13 Zoom to the Alta Bates Medical Center point marker.

Alta Bates is the closest tertiary medical care facility that has comprehensive emergency medical services. This is the likely destination for casualties occurring at the stadium event site, and hence, the probable secondary target for an IED detonation.

14 Label the FL_BAUA_GNIS data layer using the Feature_Na field.

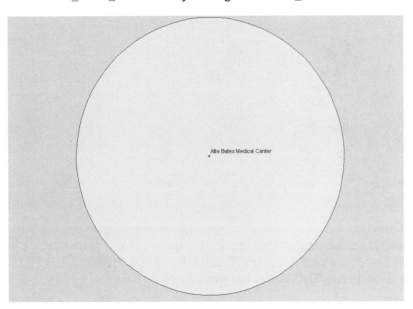

15 Add the BAUA_Transportation group data layer.

16 Drag it above the FL_BAUA_GNIS_Buffer layer, and turn on the primary, secondary, and local roads data layers.

17 Label the roads using the NAME field.

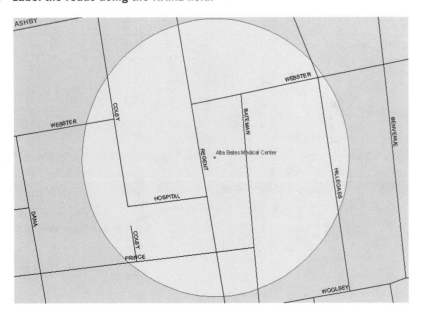

EXERCISES

You are now able to view the roads that surround the hospital site. These ingress and egress routes are to be closely monitored for suspicious activity as people come and go. Extra security precautions should be taken at the emergency entrance to the hospital facility, as this is the targeted site for the secondary IED detonation.

Secure ingress and egress routes around critical infrastructure

1   Set selectable layers to **FL_BAUA_Hospital_Buffer500.**

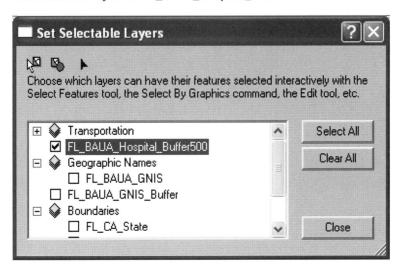

2   Use the Select tool to select only the buffer around Alta Bates Medical Center.

3   Select by location the local roads that intersect with only the selected FL_BAUA_Hospital_ Buffer500.shp.

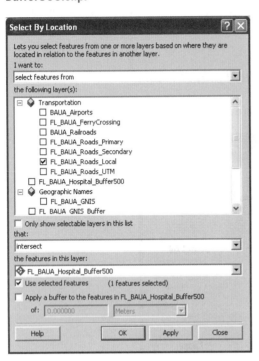

4   Click OK.

The roads that intersect with the buffer perimeter are selected and highlighted.

5   In ArcToolbox, expand the Analysis and Overlay tools, and select the Intersect tool. If necessary, enlarge the dialog box to show all options.

6   Add FL_BAUA_Hosptail_Buffer500 and FL_BAUA_Roads_Local as the Input Features; Specify the Output Feature Class **\ESRIPress\GISTHS\MYGISTHS_Work\FL_BAUA_Roads_Local_Intersec.shp**; and, the Output Type as **POINT**.

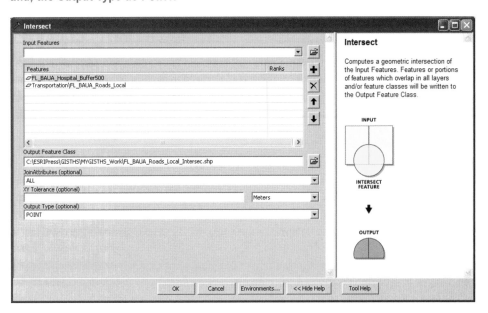

7   Click OK.

8   Close the Intersect window when processing is complete.

9   Clear selected features.

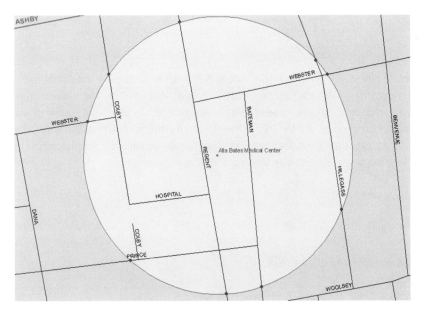

A new point feature is added to the map containing a point marker for each location where the road and buffer intersect. These positions are the optimal locations for increased security to monitor ingress and egress to the hospital site.

10 Save the ArcMap document as **FL_GISTHS_C5E3a.mxd**, with FL being your initials, in **\ESRIPress \GISTHS\MYGISTHS_Work.**

YOUR TURN

Select another hospital site within the three-mile buffer around the stadium and perform this analysis to identify the optimal locations for increased security to monitor ingress and egress to this site.

Site surveillance locations around critical infrastructure

The optimal location of surveillance positions is critical to the safety and security of critical infrastructure. Whether they are traditional closed circuit TV surveillance systems or newer enhanced IP-video security applications, where such devices are deployed determines the degree of success in combating threats to security.

In our 3D world, there are many obstacles that may obstruct our view across the landscape. Such obstacles pose logistics challenges to security personnel. A 2D map displays data as if you were flying overhead, showing where features, such as roads, landmarks, mountains, and vegetation are located. This perspective, however, doesn't take into account how the height of these features and the relief of the land can obscure a view or line of sight from one observation point to another.

A set of tools found in the 3D Analyst extension further enhances the visualization and analytical functionality of ArcMap. Viewing and analyzing a site in 3D enables a homeland security planner to determine what is visible from a chosen location near a targeted site in order to position surveillance systems at the most optimal locations.

Generate hillshade

The stadium site and immediate environs require high-level site surveillance to secure the area from the potential staging of this IED threat. The hillshade function adds a shaded relief effect to the surface layer to show how the landscape at a particular location is shaded at a certain time of day at a certain time of year. The default setting for this tool is an azimuth of 315°, or northwest, at an altitude of 45°, or halfway between the horizon and directly overhead. Generating a hillshade surface set for the actual time and date of an event is very useful in visualizing how the landscape relief is shaded at that particular time.

1 Open a new ArcMap document.

2 From the Main menu, click Tools, Extensions.

3 Click the box beside 3D Analyst and click Close.

4 From the Main menu, click View, Toolbars, and select 3D Analyst.

5 Dock this menu bar in a spot at the top or side of the map display.

6 Add the elevation group layer from **\ESRIPress\GISTHS\MYGISTHS_Work.**

7 From the 3DAnalyst menu, select Surface Analysis, Hillshade.

8 Select the elevation surface layer, **FL_BAUA_DEM**, as the input layer.

9 Type **270** as the azimuth, and **39** as the altitude.[16]

10 Check to Model shadows, and accept the Z factor and output cell size.

11 Save the output raster as **FL_HS613074PM**, with **FL** being your initials, in **\ESRIPress\GISTHS \MYGISTHS_Work**.

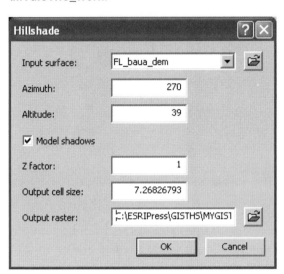

12 Click OK.

A hillshade surface is added to the map display. The darker gray areas indicate those places that are shaded at this particular date and time at this location.

13 Add the Orthoimagery data layer and drag it above the hillshade layer.

14 Zoom to the extent of the Orthoimagery data layer and increase its transparency to 60%.

The hillside area along the ingress/egress route to the northeast of the stadium becomes shaded at this time of day. As the day of the event progresses, areas such as this will become shaded and obstructed from view, and may require addition security surveillance.

YOUR TURN

Go to the U.S. Naval Observatory Astronomical Applications Department Web site (http://aa.usno.navy.mil) to find out the azimuth and altitude of the city of Berkeley two hours later at 6 p.m. on the same day, June 13, 2007. Generate a new hillshade surface under the orthoimage to see which areas are now obscured from direct sunlight and may require additional security surveillance. Plotting a temporal series of these maps will assist homeland security operations personnel to manage and deploy security forces at these locations over the course of the event.

15 Save the ArcMap document as **FL_GISTHS_C5E3b.mxd**, with FL being your initials, in **\ESRIPress \GISTHS\MYGISTHS_Work**.

Create new shapefile of surveillance points in ArcCatalog

1 Click the ArcCatalog button to launch ArcCatalog.

2 Navigate to your **\ESRIPress\GISTHS\MYGISTHS_Work** folder.

3 Click the File menu, and select New > Shapefile.

4 Name the new shapefile **FL_surveil_pts**, with FL being your initials.

5   Click the Edit button to open the Spatial Reference Properties box.

6   Click Import and navigate to your elevation surface layer, FL_BAUA_DEM, in the **\ESRIPress \GISTHS\MYGISTHS_Work** folder to import its x,y coordinate system.

7   Click OK.

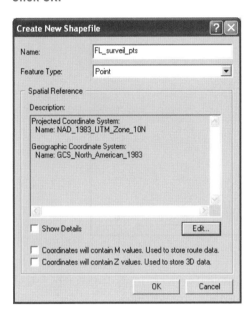

8   Click OK to close the Create New Shapefile box.

Add new shapefile to map document

1   Add this new FL_surveil_pts shapefile to the map document.

2   Change the symbol marker to a 12-point red triangle.

3   Reduce the transparency of the orthoimagery data layer to 0%.

4   If the Editor toolbar is not visible, click View from the Main Menu, and select Toolbars, and Editor.

5   Dock this menu bar in a spot at the top or side of the map display.

6   Click the Editor drop-down menu and select Start Editing.

7   If the Start Editing window appears, select the source folder **\ESRIPress\GISTHS\MYGISTHS_Work** where the FL_surveil_pts shapefile is located.

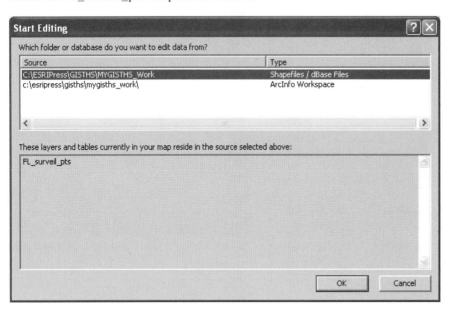

The FL_surveil_pts shapefile is listed in the box and available for editing. If this window does not appear, it means that only this one data layer is available for editing.

8   Click OK.

9   Click the Sketch Tool  and click once to place a point at each open clearing on the hillside to the north of the stadium along ingress and egress routes. (Note: Your image extent may be slightly different. If all clearing areas shown in this graphic are not available, include areas in your image extent that are.)

10   Click the Editor drop-down menu and select Stop Editing.

11 Click Yes to save your edits.

12 Save the ArcMap document as **FL_GISTHS_C5E3c.mxd**, with FL being your initials, in **\ESRIPress \GISTHS\MYGISTHS_Work**.

Generate viewshed

Viewshed analysis identifies locations that can be seen from one or more observation points in an area. It takes into account the relief of the landscape, as well as the height or offset of the observer and target. Homeland security planners can employ viewshed analysis to determine which surveillance sites around critical infrastructure provide the greatest visible coverage. This can ensure that no suspect activity is occurring out of sight of security personnel.

1 From the 3D Analyst toolbar, select Options.

2 Click the Extent tab and select the elevation surface layer, FL_BAUA_DEM, as the Analysis extent.

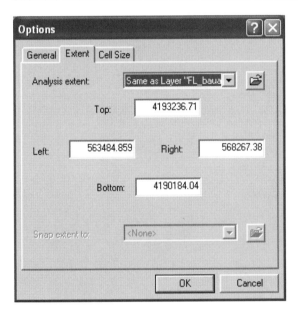

3 Click OK.

4 Zoom to the extent of the elevation surface layer, and turn off the Orthoimagery layer.

5 From the 3D Analyst menu, select Surface Analysis, Viewshed.

6 Select the elevation surface layer, FL_BAUA_DEM, as the input surface, and FL_surveil_pts as the observer points.

The curvature of the earth decreases the height of a feature eight meters for each 10 kilometers in distance. Because this is a large scale map of a small area, it is not necessary to use the earth curvature in this viewshed analysis.

7   Accept the Z factor and output cell size, and save the output raster as **FL_Viewshed0**, with FL being your initials, in **\ESRIPress\GISTHS\MYGISTHS_Work**.

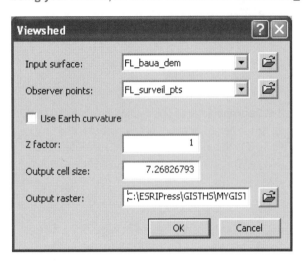

8   Click OK.

The viewshed map is drawn to the display. The pink areas indicate the locations that are not visible from these surveillance points, and the green areas indicate those that are.

9   Increase the transparency of the viewshed layer to 50%.

10   Turn on the orthoimagery layer and zoom to its extent.

Note there are a few locations along the ingress/egress routes that are not visible from these surveillance points. Increasing the source height of these surveillance points by erecting observation towers will increase the viewshed range of these sites.

11   Right-click FL_surveil_pts and open the attribute table.

12   Click the Options button and select Add Field.

Note: If you get a message that the table/feature class is in use by another application/user, close ArcCatalog.

13   Enter **OFFSETA** as the field name, and accept Short Integer as the type.

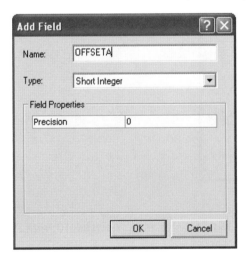

14   Click OK.

15 Click the Editor drop-down menu and select Start Editing.

16 If the Start Editing window appears, select the source folder **\ESRIPress\GISTHS\MYGISTHS_Work** where the FL_surveil_pts shapefile is located.

17 Edit the OFFSETA field for each point from 0 to 10 to account for an observation tower height of approximately 30 feet.[17]

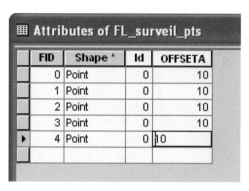

18 Stop editing, save your edits, and close the Attributes of FL_surveil_pts table.

19 From the 3D Analyst menu, select Surface Analysis, Viewshed. You will run a revised viewshed analysis using the height of the observation towers as the source height.

20 Select the elevation surface layer, **FL_BAUA_DEM**, as the input surface, and FL_surveil_pts as the observer points.

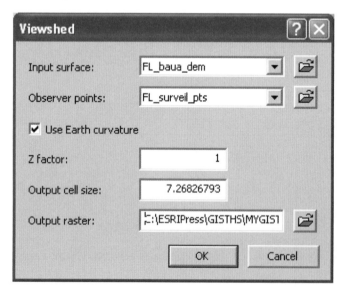

21 Accept the Z factor and output cell size, and save the output raster as **FL_Viewshed10**, with FL being your initials, in **\ESRIPress\GISTHS\MYGISTHS_Work**.

22 Click OK.

The new semitransparent viewshed layer is draped over the original viewshed layer.

23 **Increase the transparency of the new viewshed layer to 50%.**

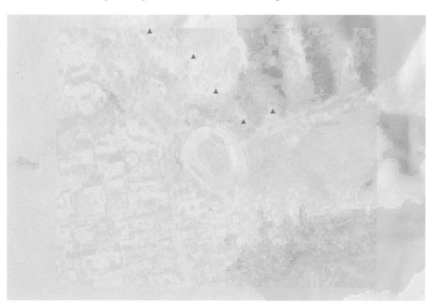

Note that more of the locations along the ingress/egress routes that were not visible from these security points when they were at ground level are now visible. A viewshed analysis such as this enhances homeland security planners' ability to site the optimal locations for effective surveillance systems designed to protect critical infrastructure from targeted threats.

Enhanced viewshed analysis can also include embedding 3D building features into the surface, such as buildings and overhead structures. By doing so, the heights of these structures can also be considered when determining the viewshed from designated observation positions.

YOUR **TURN**

Create a new shapefile of alternate surveillance points at other locations along the hillside. Run a revised viewshed analysis to see how the range of view from these locations differs from the original surveillance sites.

24 Save the ArcMap document as **FL_GISTHS_C5E3d.mxd,** with FL being your initials, in **\ESRIPress \GISTHS\MYGISTHS_Work.**

Create line of sight

An alternate method for determining the greatest field of view for optimal surveillance coverage includes the creation of line-of-sight polylines. A line of sight is a graphic line between two points on a surface that shows where along the line the view is obstructed.[18] This tool is available in the 3D Analyst extension, and can also be added to a 3D ArcScene.

1   Turn off all layers except the surveillance points, orthoimage, and FL_Viewshed10 surface layer.

Draping the semitransparent viewshed surface over the orthoimage reveals the areas that are obstructed from the view of the surveillance positions.

2   Use the Zoom In tool to zoom to the northern edge of the stadium where the ground is obstructed from view.

3   Select the elevation surface layer, FL_BAUA_DEM, from the 3D Analyst toolbar.

4   Click the Create Line of Sight tool ⬦→ on the 3D Analyst toolbar.

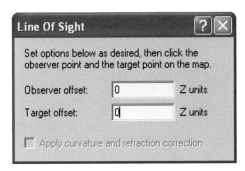

The Line of Sight dialog box appears. These settings enable you to offset the height of the surveillance unit and target to account for their height above the surface.

5   Enter **20** as the observer offset to account for the stadium height of approximately 65 feet.

6   Enter **1** as the target height to enable viewing of objects low to the ground.

The check box to apply curvature and refraction correction is not available as the curvature of the earth is negligible at this large map scale.

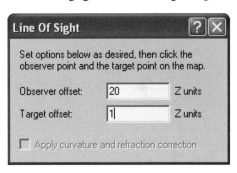

7   Click at the edge of the stadium to the north where the ground is obstructed from view from the observation towers.

Note: Your map display may be slightly different.

8   Drag the cursor out to the road area that is also obscured from view, and release the mouse.

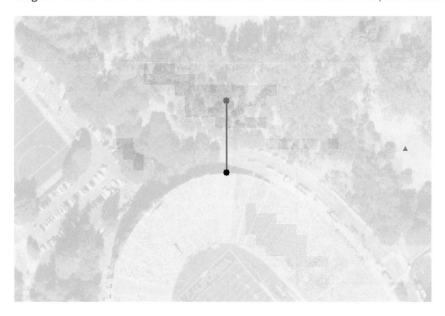

A black dot is drawn at the stadium edge where the surveillance unit will be deployed. A green polyline is drawn across to the road where a green dot is placed. This means that the view from the surveillance unit to the road will not be obstructed by terrain relief. Other 3D map objects, if available, can be added to the surface, such as trees and buildings, to further enhance the line-of-sight capabilities. However, this exercise is limited to the terrain relief of the landscape.

9 Continue clicking around the perimeter of the stadium to test the line of sight of other potential surveillance positions. Zoom out if necessary to broaden the extent of the line of sight.

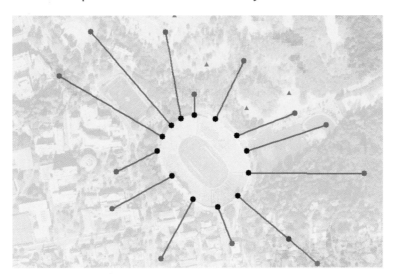

Most positions from the top perimeter of the stadium structure will have a clear line of sight to the target locations on the ground in all directions. There are, however, a few lines of sight that are obstructed from full view. The red polyline shows those areas not visible from the observer point; the blue dot indicates where the obstruction is located; and the red dot indicates that the target is not within view. These observation positions would not be suitable to view the further southeastern extent of the stadium environs.

10 Set the observer offset to 15 to account for the 50-foot height of the round tower at the western edge of the orthoimage.

11 Create a set of line-of-sight polylines from that tower to the parking lot and tennis court areas.

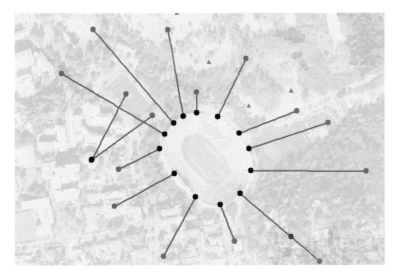

You now have a series of observation points where surveillance cameras can be deployed to oversee the operations and activities before, during, and after an event.

12 Save the ArcMap document as **FL_GISTHS_C5E3e.mxd**, with FL being your initials, in **\ESRIPress \GISTHS\MYGISTHS_Work**.

Create line-of-sight profile graph

ArcMap enables you to view the line of sight not only from above, but also as a profile graph. This profile view provides a perspective of the line of sight as if you were observing it from the ground. It offers a better look at where the view may be clear or obstructed.

1   Use the Select Elements tool ![cursor] to click on one of the lines of sight where there is a view obstruction.

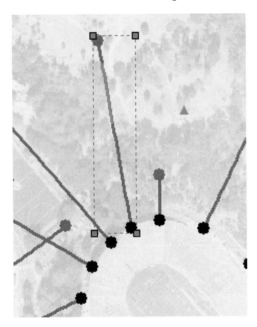

A set of four blue handles is placed at the corners of the selection box around the line element.

2   Click the Create Profile Graph tool ![icon] on the 3D Analyst toolbar.

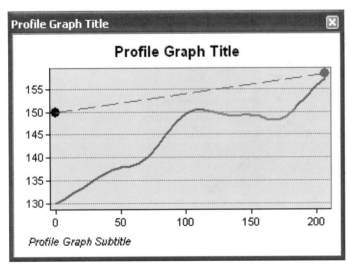

A profile graph is created of this line of sight, showing the extent of the line of sight as it traverses across the landscape. Areas in full view are drawn in green, while sites that are obstructed from view are drawn in red. You can change the properties of this graph to customize it to the specific analysis you are doing.

3   Right-click the graph and select Properties.

4   Click the Appearance tab.

5   Change the Title to: **Memorial Stadium Observation Point**; and the Footer to **Line of Sight Analysis**.

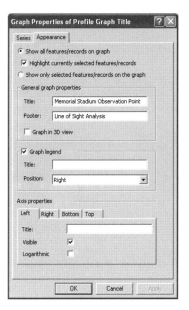

6   Click OK.

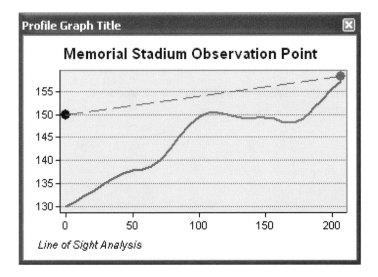

The graph is updated to show the updated title and footer. The graph can be printed, duplicated, copied as a graphic, added to a layout, saved, or exported for use in a layout or report.

7   Right-click the graph and select Export.

8   Select as JPEG for the Format list in the Export Dialog window.

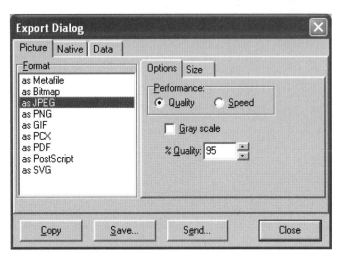

9   Click Save and save it as **FL_MemStad_LOS**, with FL being your initials, in **\ESRIPress\GISTHS \MYGISTHS_Work.**

10 Click Close and close the profile graph window.

This profile graph can now be used to enhance the line-of-sight analysis to assist homeland security operations personnel as they test various locations around the stadium for possible surveillance points.

View 3D line of sight

Lines of sight, and observer, obstruction, and target points can be viewed in 3D using the ArcScene viewer. This view adds a more realistic perspective to the map to assist homeland security operations personnel in planning and implementing surveillance measures.

1   Click the ArcScene button ![icon] from the 3D Analyst toolbar to launch ArcScene.

2   Add the elevation surface layer, FL_BAUA_DEM, to the ArcScene display.

3   Right-click FL_BAUA_DEM in the Table of Contents and select Properties.

4   Click the Base Heights tab.

5   Click the radio button to obtain heights for layer from surface. If the elevation surface layer is not already prefilled, navigate to **\ESRIPress\GISTHS\GISTHS_C3\Downloaded_Data_C3\FL_BAUA_DEM** to locate this layer.

6   Click OK.

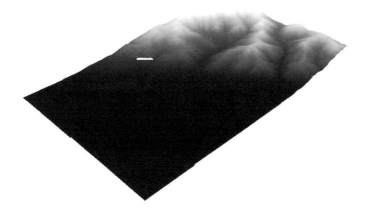

The FL_BAUA_DEM layer now appears as a 3D surface layer.

7   Add the Orthoimagery data layer to ArcScene, and drag it above FL_BAUA_DEM in the Table of Contents.

8   Right-click the Orthoimagery data layer and select Properties.

9   Click the Base Heights tab.

10 Click the radio button to obtain heights for layer from surface. The elevation surface layer is already prefilled.

11 Enter 30 to offset the Orthoimage data layer above the elevation surface layer.

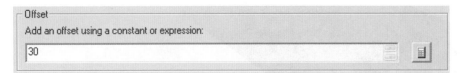

12 Click OK.

The orthoimage is draped over the elevation surface layer.

13 Use the Zoom In/Out tool on the ArcScene toolbar to zoom in to the orthoimage.

14 Use the Navigate tool ⊕ on the ArcScene toolbar to traverse the 3D surface from different perspectives.

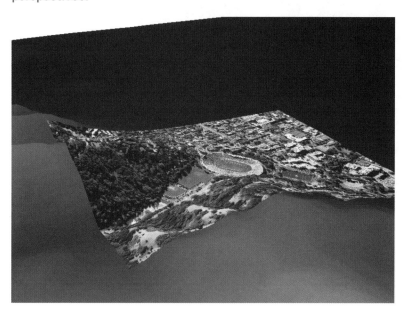

15 Return to ArcMap and click the Select Elements tool ▶ from the toolbar.

16 Click on one of the line-of-sight polylines to place a box with graphics handles at the four corners.

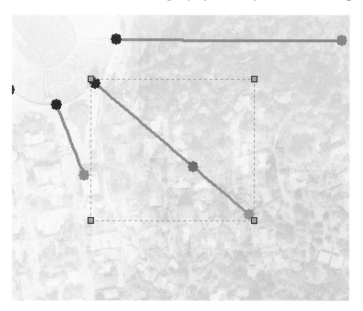

17 From the Main Menu, click Edit and select Copy.

18 Return to ArcScene.

19 From the Main Menu, click Edit and select Paste.

20 Turn off the Orthoimagery layer to reveal the line of sight polyline.

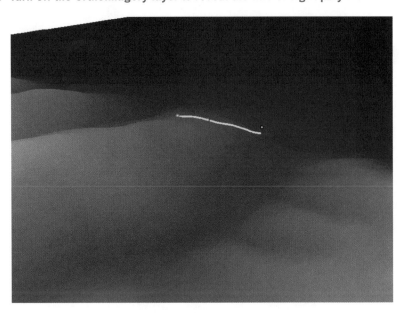

Since the line-of-sight polyline is a 2D graphic feature, it rests directly upon the elevation surface layer.

21 Use the Select Elements tool to click on the line-of-sight polyline to unselect it.

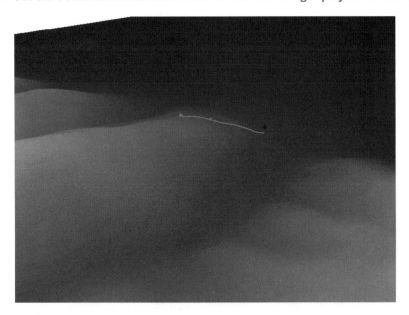

22 Turn on and zoom to the extent of the Orthoimagery layer.

23 Right-click the Orthoimagery layer and select Properties.

24 Click the Display tab and set the transparency to 60%.

Contrast:	0	%
Brightness:	0	%
Transparency:	60	%

25 Click OK.

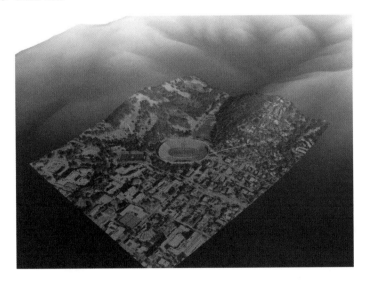

The line of sight polyline is now visible through the Orthoimagery data layer, revealing how the terrain relief obstructs the range of view from that designated surveillance position.

26 Copy and paste additional line-of-sight polylines to the ArcScene to show how they extend over the surface.

27 Save the ArcScene document as **FL_GISTHS_C5E3.sxd**, with FL being your initials, in **\ESRIPress \GISTHS\MYGISTHS_Work**.

28 Close ArcScene.

29 Return to ArcMap and save the ArcMap document as **FL_GISTHS_C5E3f.mxd** in **\ESRIPress \GISTHS\MYGISTHS_Work**.

30 Close the map document.

Chapter 5 (5.4-5.6)

Analyzing data for homeland security planning and operations—*Protect*

DHS National Planning Scenario 6: *Chemical attack—toxic industrial chemicals*

MISSION AREA: *PREVENT*
Visibly increasing security and apprehension potential at the site before and during the attack.

MISSION AREA: *PROTECT*
DHS Target Capability: *Citizen protection*
DHS Task: *Pro.C.2.1.9 Provide public safety—develop protection plans for special needs populations*

5.4	**Prepare protect scenario map 2**
5.5	**Locate impacted population** Download census data from ESRI data portal Define census shapefile coordinate system Add census shapefile to map document Join demographic data to census geography shapefile Create a population choropleth map Compute the impacted population in the plume area Locate and compute special-needs population in the plume area
5.6	**Protect impacted population** Identify assembly points outside plume area

MISSION AREA: *RESPOND*
Alerts, activation and notification, traffic and access control, protection of special populations, resource support, requests for assistance, and public information.

MISSION AREA: *RECOVER*
Decontamination of concentrated areas; disposal of contaminated wastes; environmental testing; repair of destroyed/damaged facilities; and public information activities.

Exercise 5.4
Prepare protect scenario map 2

DHS National Planning Scenario 6:

Chemical attack—toxic industrial chemicals

DHS Target Capability:

Citizen protection: Evacuation and/or in-place protection

DHS Task:

Pro.C.2.1.9 provide public safety—develop protection plans for special needs populations

DHS National Planning Scenario 6: Chemical attack—toxic industrial chemicals is described as follows:

> In this scenario, terrorists from the Universal Adversary (UA) land in several helicopters at fixed-facility petroleum refineries. They quickly launch rocket-propelled grenades (RPGs) and plant improvised explosive devices (IEDs) before reboarding and departing, resulting in major fires. At the same time, multiple cargo containers at a nearby port explode aboard or near several cargo ships with resulting fires. Two of the ships contain flammable liquids or solids. The wind is headed in the north-northeast direction, and there is a large, heavy plume of smoke drifting into heavily populated areas, releasing various metals into the air. One of the burning ships in the port contains resins and coatings including isocyanates, nitriles, and epoxy resins. Some IEDs are set for delayed detonation. Casualties occur onsite due to explosive blast and fragmentation, fire, and vapor/liquid exposure to the toxic industrial chemical (TIC). Downwind casualties occur due to vapor exposure. [19]

The scenario profile includes a synopsis of how the *Protect* Mission Area is to be activated in response to this threat. It clearly stipulates that the protection of the public includes the evacuation and/or sheltering of downwind populations.[20]

This GIS exercise focuses on locating the heavily populated areas where the toxic plume could potentially be released. In addition, you will locate special needs populations, such as the elderly who may require additional assistance during evacuation or in-place protection. Knowing beforehand where additional protective resources are to be expended ensures people with special needs will not be overlooked and will be secure during an event. Selected facilities outside of the plume area are also identified to provide assembling points for evacuated residents.

1   **Open a new ArcMap document.**

2   **Add the FL_BAUA_MEDS_Boundaries, FL_BAUA_MEDS_Hydrography, and FL_BAUA_MEDS_ Transportation group layers from the \ESRIPress\GISTHS\MYGISTHS_Work folder.**

3   **Order and turn on the NHD_Area, primary, secondary and local roads, and FL_BAUA_Counties data layers.**

4   **Add the Petrol_Tanks and PLUME_COBALT_HYDROCARBONYL layer files from the \ESRIPress \GISTHS\GISTHS_C5 folder.**

5   **Drag the Plume layer to the top of the Table of Contents.**

6 Zoom to the extent of the **PLUME_COBALT_HYDROCARBONYL** data layer, and increase its transparency to 50%.

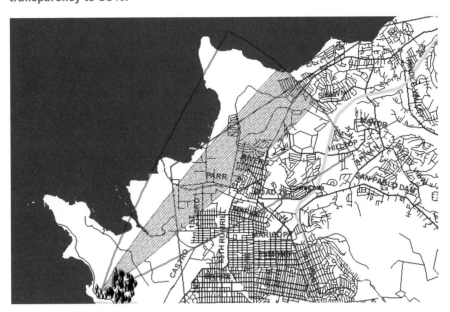

This map displays a refinery field in Richmond, California, along the inner coast harbor of the San Francisco Bay. This site could be a likely target for an attack as described in the national planning scenario above. If one of the oil tanks or cargo containers at this location were attacked, a toxic industrial chemical could, given the wind conditions at the time of the attack, be released and carried north to northeast in the direction of a heavily populated community.

This plume data layer was generated using the program Areal Locations of Hazardous Atmospheres (ALOHA), part of an emergency response planning tool, Computer-aided Management of Environmental Operations (CAMEO), created by a joint effort between the U.S. Department of Environmental Protection and the National Oceanic and Atmospheric Administration.[21] The parameters used to define the plume include the coordinate locations of the site and those stipulated in the planning scenario details: the chemical cobalt hydrocarbonyl, wind at 5 knots coming from the southwest, air temperature of 70 degrees, and relative humidity of 35 percent. The threat zone or plume footprint generated by the ALOHA software is imported into ArcMap via the ArcMap ALOHA Import Tool, and saved as a shapefile. The shapefile contains two attributes: plume footprint, which is the extent of the area directly impacted by the toxic release, and the wind direction confidence extent, which covers the area that the wind and plume may shift given the weather conditions entered into the model.

A field operations tool that is worth noting at this point is one that at the time of this writing is currently in development by an NSF funded joint collaboration between ESRI and Jackson State University. The All Hazards Emergency Operations Management System (ALLHAZ) is an application that will provide all field operations personnel with a standardized, scalable, geospatially enabled tool that they can use to assist in planning for, mitigating, responding to, and recovering from all hazards of all sizes.

It is designed to:

> "implement key components of DHS' National Preparedness Goal at the local level and to provide a Common Operation Picture employing geospatial displays and models. ALLHAZ leverages current project, existing data and resources, heterogeneous data sources and is targeted for use by Responders teams. ALLHAZ is scaleable to support routine operations and large scale disaster." [22]

(Roadblock.wmv, ESRI, David Kehrlein, Elizabeth Matlack)

In the case of a toxic hazard, as addressed in this scenario, the ALLHAZ tool enables a field response operator to identify the point source of the hazard and then generate the ALOHA plume layer via a wireless handheld remote device that is part of a common operating picture (COP) system shared by other personnel involved in planning and facilitating response and recovery operations throughout a common area.

Such devices will greatly enhance future onsite response and recovery capabilities of homeland security personnel at the command center and in the field as more data is presented and managed in real time via an integrated and shared common operating environment. Geospatial analysis in a desktop or server environment can then be shared with data generated at the mobile level to identify impacted populations, locate evacuation and shelter facilities, and then pinpoint the optimal location of road blocks along ingress and egress routes in and around the plume area.

6   Save the ArcMap document as **FL_GISTHS_C5E4.mxd**, with FL being your initials, in **\ESRIPress \GISTHS\MYGISTHS_Work**.

Exercise 5.5
Locate impacted population

Census data can help homeland security planners prepare tactics to inform and protect a potentially impacted population in the event of an evacuation or in-place protection order. The U.S. Census Bureau prepares geospatial data available via ESRI's Geography Network data portal.

Download census data from ESRI data portal

1   Navigate your Internet browser to www.esri.com/data/data_portals.html.

2   Click Census 2000 TIGER/Line Data.

3   Click Preview and Download under the Free Download heading in the left column.

4   Select California and click Submit Selection.

5   Select Contra Costa and click Submit Selection.

6   Put a check mark in the Census Blocks 2000 box.

7   Click Proceed to Download.

8   Click Download and save the zip file in **\ESRIPress\GISTHS\MYGISTHS_Work**.

9   When the download is complete, click the back button to return to the data list.

10 Scroll down and click the Census Block Demographics (SF1) link.

11 Click Download File and save the zip file in **\ESRIPress\GISTHS\MYGISTHS_Work**.

12 Navigate to the zip files and extract the two zip files to four files in **\ESRIPress\GISTHS \MYGISTHS_Work**.

   • tgr06013blk00.dbf
   • tgr06013blk00.shp
   • tgr06013blk00.shx
   • tgr06000sf1blk.dbf

   Note: If you have limited Internet connectivity, these files are available locally at \ESRIPress\GISTHS\GISTHS_C5\Downloaded_Data_C5. Copy them into your \ESRIPress\GISTHS\MYGISTHS_Work folder before you begin working with them.

Define census shapefile coordinate system

1   In ArcCatalog, right-click tgr06013blk00.shp, and select Properties.

2   Click the XY Coordinate System tab.

EXERCISES

3   Click Select to define the coordinate system as Geographic Coordinate System North America North American Datum 1983.

4   Click OK.

Add census shapefile to map document

1   Add tgr06013blk00.shp to the map document, and drag it below all data layers except the boundaries group layer.

2   Zoom to the extent of the plume data layer and turn off the Transportation group data layers.

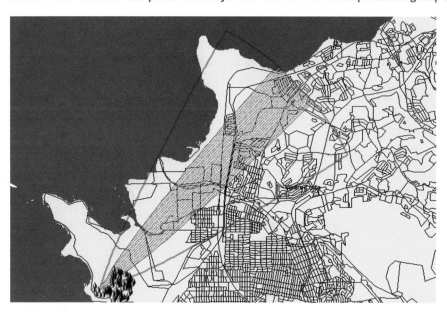

Each polygon represents a census block, which is a subdivision of a census tract. It is the smallest geographic unit for which the Census Bureau tabulates 100 percent data. Many census blocks correspond to individual city blocks bounded by streets but blocks, especially in rural areas, may include many square miles and may have some boundaries that are not streets.[23]

3 Open the tgr06013blk00 attribute table.

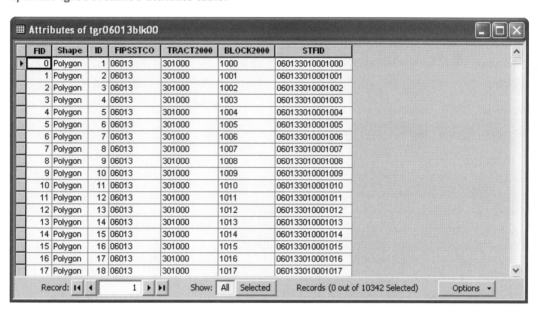

This table only contains the attribute information about the geography of these census block groups. There is no demographic information included in this file. The demographic information is contained in the tgr06000sf1grp.dbf file that was downloaded with this shapefile from the ESRI data portal.

4 Add tgr06000sf1blk.dbf to the map document and open it.

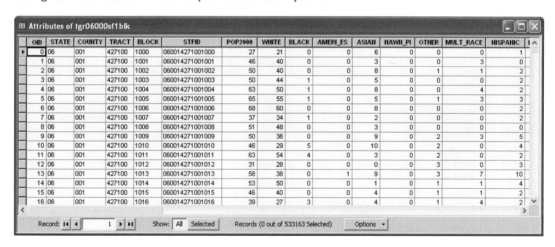

The demographic data is contained in this table. It can be joined to the census geography shapefile to compute the population located in the plume area.

5 Close the Attributes of tgr06000sf1blk table.

EXERCISES

Join demographic data to census geography shapefile

1   Join the tgr06000sf1blk.dbf table to the tgr06013blk00 shapefile using STFID as the join field.

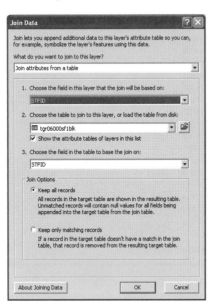

2   Click OK, and Yes to create an index, if asked.

3   Scroll to the POP2000 field in the Attributes of tgr06013blk00 table.

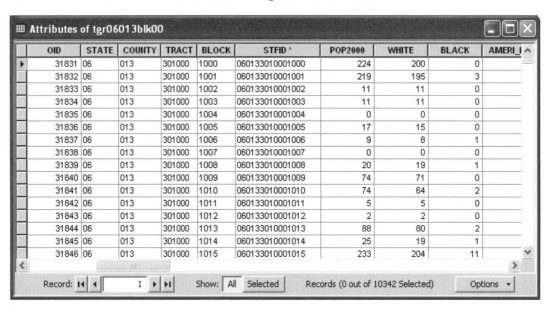

OID	STATE	COUNTY	TRACT	BLOCK	STFID *	POP2000	WHITE	BLACK	AMERI_I
31831	06	013	301000	1000	060133010001000	224	200	0	
31832	06	013	301000	1001	060133010001001	219	195	3	
31833	06	013	301000	1002	060133010001002	11	11	0	
31834	06	013	301000	1003	060133010001003	11	11	0	
31835	06	013	301000	1004	060133010001004	0	0	0	
31836	06	013	301000	1005	060133010001005	17	15	0	
31837	06	013	301000	1006	060133010001006	9	8	1	
31838	06	013	301000	1007	060133010001007	0	0	0	
31839	06	013	301000	1008	060133010001008	20	19	1	
31840	06	013	301000	1009	060133010001009	74	71	0	
31841	06	013	301000	1010	060133010001010	74	64	2	
31842	06	013	301000	1011	060133010001011	5	5	0	
31843	06	013	301000	1012	060133010001012	2	2	0	
31844	06	013	301000	1013	060133010001013	88	80	2	
31845	06	013	301000	1014	060133010001014	25	19	1	
31846	06	013	301000	1015	060133010001015	233	204	11	

Record: |◀| |◀| 1 |▶| |▶| Show: All Selected   Records (0 out of 10342 Selected)   Options ▼

The demographic data is now joined to the census geography shapefile and can be mapped to show the population within the plume area.

4   Close the Attributes of tgr06013blk00 table.

5   Export the tgr06013blk00 shapefile using the coordinate system of the data frame, and save it as **FL_BAUA_CensusBlocks.shp**, with FL being your initials, in **\ESRIPress\GISTHS\MYGISTHS_Work**.

6   Click Yes to add it to the map as a layer.

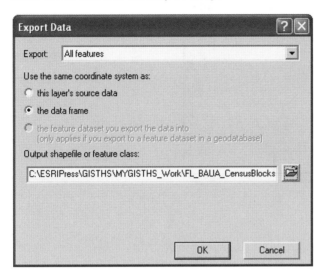

7   Remove tgr06013blk00.shp and tgr06000sf1blk.dbf from the map document.

8   Drag the new FL_BAUA_CensusBlocks.shp file below all data layers except the boundaries group data layer.

Create a population choropleth map

1   Right-click FL_BAUA_CensusBlocks and select Properties.

2   Click the Symbology tab and change the symbology to Quantities using POP2000 as the Value field.

3   Click OK to confirm that the maximum sample size has been reached.

4   Click the Classify button.

5   Click the Sampling button.

6   The Attributes of tgr06013blk00 table contains 10,342 records. The maximum sample size is set to a default of 10,000. Increase this amount to 11,000.

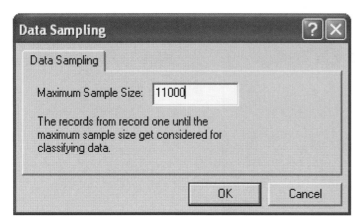

7   Click OK, twice.

8   Choose a graduated color ramp to show the range of population.

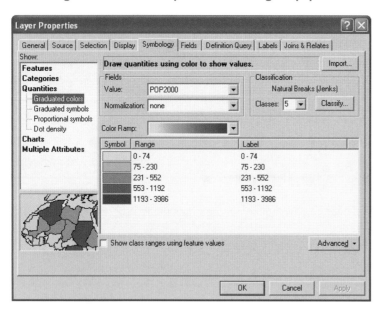

The classification symbology is now set for this data layer.

9   Click OK.

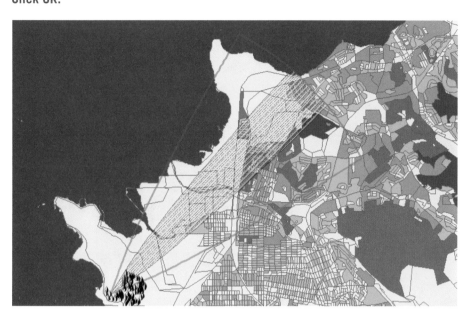

The map now shows the total population of each census block in the county, and can be viewed below the extent of the plume.

Compute the impacted population in the plume area

1   Select by Location features from **FL_BAUA_CensusBlocks** that intersect with the features of the **PLUME_COBALT_HYDROCARBONYL** data layer.

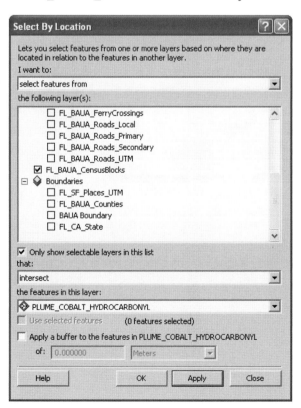

2   Click Apply and Close.

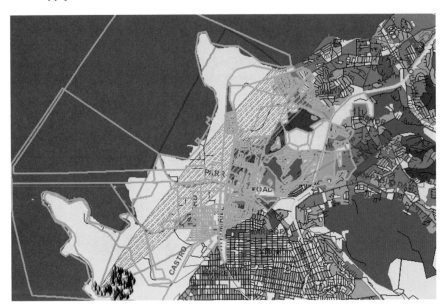

Each of the census blocks impacted by the plume is selected. Note that only part of some census blocks are within the plume extent but the entire polygon is selected.

3   Open the Attributes of FL_BAUA_CensusBlocks table.

4   Click Selected to show only those census block groups that are in the plume area.

5   Right-click the POP2000 field heading and select Statistics.

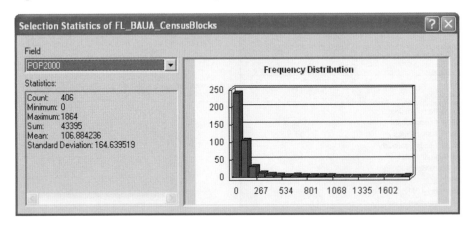

The Statistics window shows 43,395 people living in the census blocks that are completely or partially within the plume area.

6   Close the Statistics window.

Compute and locate special-needs population in the plume area

1 Right-click the Age_65_UP field heading and select Statistics.

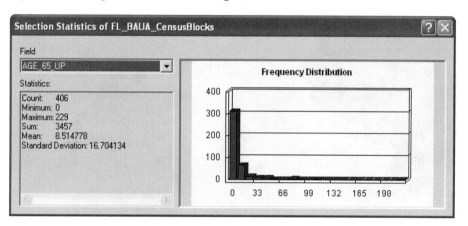

The Statistics window shows 3,457 people aged 65 and older living in the census blocks that are completely or partially within the plume area.

2 Close the Statistics window and the Attributes of FL_BAUA_CensusBlocks table.

3 Right-click FL_BAUA_CensusBlocks and select Properties.

4 Click the Symbology tab and change the symbology to Quantities using AGE_65_UP as the Value field.

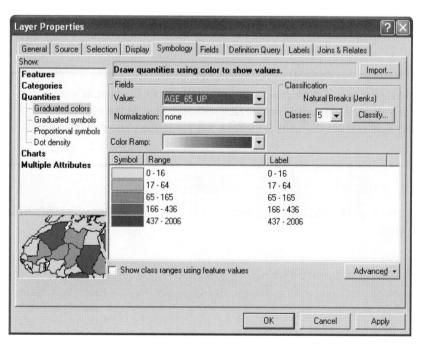

5 Click OK.

6   Clear selected features.

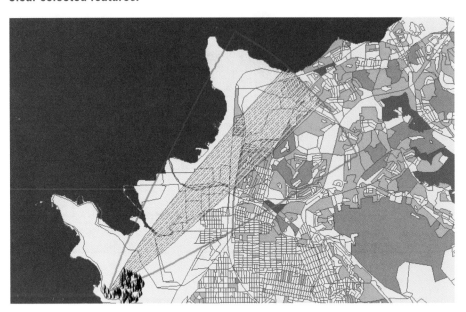

This map now shows the distribution of elderly residents in and around the impacted plume area. This map, coupled with the statistics tallying the elderly population, can be very useful to local Emergency Operations Center personnel who can plan for the residents' safe and effective evacuation or in-place protection.[24] Considerations include knowing where to allocate resources to transport elderly residents from nursing care facilities; contacting next of kin; and tracking medications, prescriptions, and special equipment like wheelchairs, canes, hearing aids, and walkers.

YOUR TURN

Map another field from the census block group table that may require consideration when designing a safe and effective evacuation and in-place protection plan. These may include statistics about the number of housing units in the area, average family size, or children under the age of 5.

7   Save the ArcMap document as **FL_GISTHS_C5E5.mxd**, with FL being your initials, in **\ESRIPress \GISTHS\MYGISTHS_Work**.

Exercise 5.6
Protect impacted population

Now that you have located and identified the population in the area potentially affected by a toxic plume, you can use GIS to design protective measures that would ensure the safety of residents either evacuated or protected in-place.

Successful evacuation planning includes many critical components,[25] including designating assembly points adjacent to the impacted area where evacuated residents can get instructions on where and how to proceed. These sites include parks, schools, and large parking lots at sports facilities and shopping centers. Relocation facilities outside of the zone that can provide temporary housing include schools, churches, community centers, and hospitals and nursing homes with available beds for those with special needs.

Assembly points include safe areas and gathering sites adjacent to the impacted area where evacuated residents can get instructions on where and how to proceed. These sites include parks, schools, and large parking lots at sports facilities and shopping centers. Relocation facilities outside of the impacted zone that can provide temporary housing include schools, churches, community centers, and hospitals and nursing homes with available beds for those with special needs.

Identify assembly points outside plume area

1 **Add the BAUA_MEDS_Structures and Geographic Names group data layers, and turn on the Transportation group layer.**

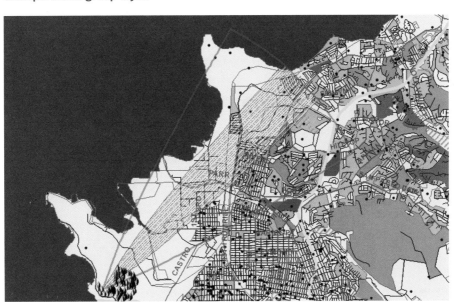

The landmarks and geographic names points are added to the map display.

EXERCISES

2 Select by location all features in the **FL_BAUA_GNIS** data layer that intersect with the **PLUME_COBALT_HYDROCARBONYL** data layer within a 1-mile buffer.

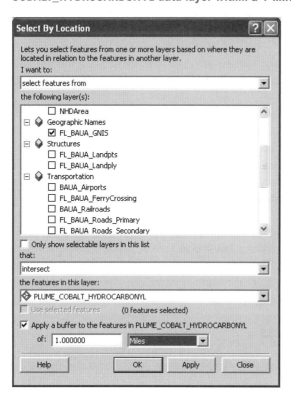

3 Select by location to remove from the currently selected features all features in the **FL_BAUA_GNIS** data layer that intersect with the plume. These features are within one mile outside of the perimeter of plume.

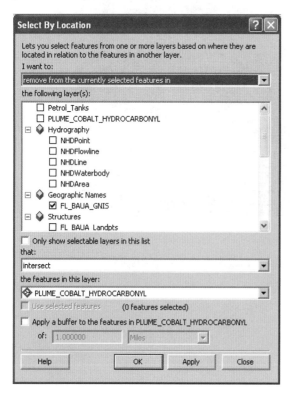

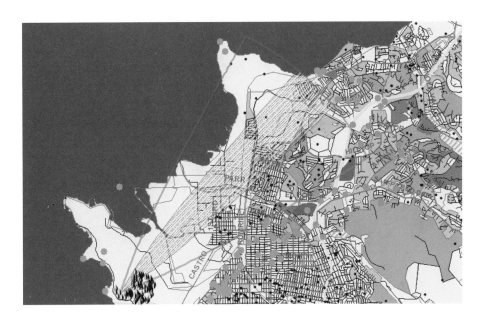

This excludes all features within the plume area from consideration as possible assembly points in the event of an evacuation order due to a toxic chemical emergency.

4   Select by attributes from the current selection all features in the **FL_BAUA_GNIS** data layer that are parks and schools.

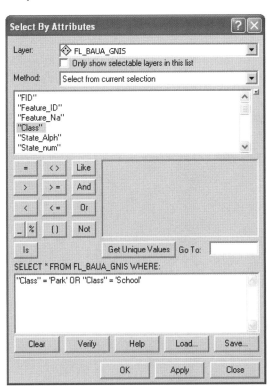

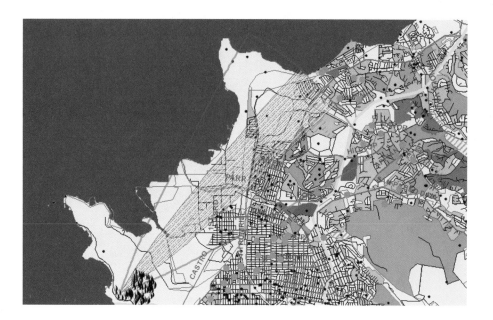

5   Export the selected features to a new shapefile using the same coordinate system as the data frame. Name the output shapefile **FL_AssemblyPts_GNIS.shp**, with FL being your initials, in **\ESRIPress\GISTHS\MYGISTHS_Work.**

6   Click Yes to add it to the map as a layer.

7   Clear selected features and turn off the FL_BAUA_GNIS data layer.

8   Turn on both Structures data layers, and select by location all features in the FL_BAUA_ LandPts data layer that intersect with the PLUME_COBALT_HYDROCARBONYL data layer within a 1-mile buffer.

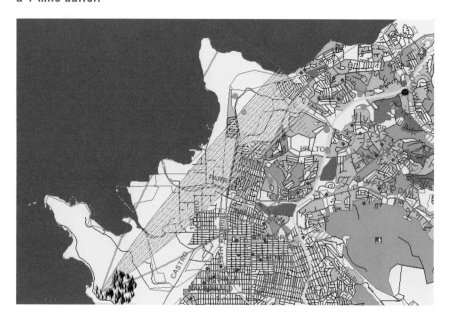

9   Select by location to remove from the currently selected features all features in the BAUA_ LandPts data layer that intersect with the PLUME_COBALT_HYDROCARBONYL data layer.

10 Open the FL_BAUA_LandPts attribute table to see that there are only four landmark points within one mile outside of the perimeter of plume.

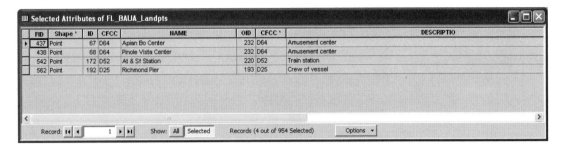

11 Select by attributes from the current selection all of the features of the FL_BAUA_LandPts data layer that are suitable for assembly points. They are described as:

"cfcc.DESCRIPTIO" = 'Amusement center' OR

"cfcc.DESCRIPTIO" = 'Campground' OR

"cfcc.DESCRIPTIO" = 'Educational institution, including academy, school, college, and university' OR

"cfcc.DESCRIPTIO" = 'Educational or religious institution' OR

"cfcc.DESCRIPTIO" = 'Golf course' OR

"cfcc.DESCRIPTIO" = 'National park or forest' OR

"cfcc.DESCRIPTIO" = 'Open space' OR

"cfcc.DESCRIPTIO" = 'Religious institution, including church, synagogue, seminary, temple, and mosque' OR

"cfcc.DESCRIPTIO" = 'Shopping center or major retail center' OR

"cfcc.DESCRIPTIO" = 'State or local park or forest'

12 Export the selected features to a new shapefile using the same coordinate system as the data frame. Name the output shapefile **FL_Assembly_Pts_LMPts.shp**, with FL being your initials, in the **\ESRIPress\GISTHS\MYGISTHS_Work** folder.

13 Close the FL_BAUA_LandPts attribute table.

14 Repeat this selection process for the FL_BAUA_LandPly data layer. Open the FL_BAUA_LandPly attribute table to see that there are 14 landmark polygon areas suitable for assembly points within one mile outside of the perimeter of plume.

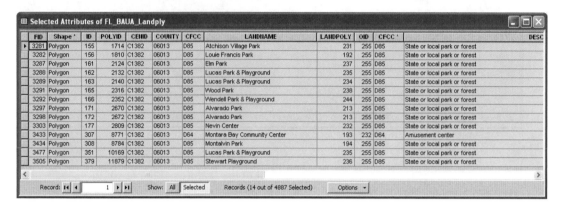

FID	Shape *	ID	POLYID	CENID	COUNTY	CFCC	LANDNAME	LANDPOLY	OID	CFCC *	DESC
3281	Polygon	155	1714	C1382	06013	D85	Atchison Village Park	231	255	D85	State or local park or forest
3282	Polygon	156	1810	C1382	06013	D85	Louie Francis Park	192	255	D85	State or local park or forest
3287	Polygon	161	2124	C1382	06013	D85	Elm Park	237	255	D85	State or local park or forest
3288	Polygon	162	2132	C1382	06013	D85	Lucas Park & Playground	235	255	D85	State or local park or forest
3289	Polygon	163	2140	C1382	06013	D85	Lucas Park & Playground	234	255	D85	State or local park or forest
3291	Polygon	165	2316	C1382	06013	D85	Wood Park	238	255	D85	State or local park or forest
3292	Polygon	166	2352	C1382	06013	D85	Wendell Park & Playground	244	255	D85	State or local park or forest
3297	Polygon	171	2670	C1382	06013	D85	Alvarado Park	213	255	D85	State or local park or forest
3298	Polygon	172	2672	C1382	06013	D85	Alvarado Park	213	255	D85	State or local park or forest
3303	Polygon	177	2809	C1382	06013	D85	Nevin Center	232	255	D85	State or local park or forest
3433	Polygon	307	8771	C1382	06013	D64	Montara Bay Community Center	193	232	D64	Amusement center
3434	Polygon	308	8784	C1382	06013	D85	Montalvin Park	194	255	D85	State or local park or forest
3477	Polygon	351	10169	C1382	06013	D85	Lucas Park & Playground	235	255	D85	State or local park or forest
3505	Polygon	379	11879	C1382	06013	D85	Stewart Playground	236	255	D85	State or local park or forest

Record: 1   Show: All | Selected   Records (14 out of 4887 Selected)   Options ▾

15 Export and name the selected features **FL_Assembly_Pts_LMPly.shp**, with FL being your initials, in the **\ESRIPress\GISTHS\MYGISTHS_Work** folder.

16 Create a new group layer named Assembly Points.

17 Drag the three assembly point data layers into the new group layer.

18 Turn off the FL_BAUA_LandPts, FL_BAUA_LandPly, and FL_BAUA_GNIS data layers.

19 Right-click the FL_AssemblyPts_GNIS data layer and select Properties.

20 Click the Symbology tab.

21 Select Categories; Unique Values; Value Field = Class; and click Add All Values.

22 Double-click the point marker for Park and click Properties from the Symbol Selector window.

23 Select Character Marker Symbol from the Type drop-down menu.

24 Select ESRI Transportation and Municipal from the Font drop-down menu.

25 Click to select the Park symbol from the symbol palette.

26 Increase the size to 14 and change the color to green.

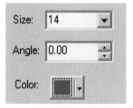

27 Click OK twice to close the Symbol Property Editor and Symbol Selector.

28 Select School 1 Size 14 from the Symbol Selector for the symbol for School.

29 Uncheck the box for <all other values>.

30 Click OK.

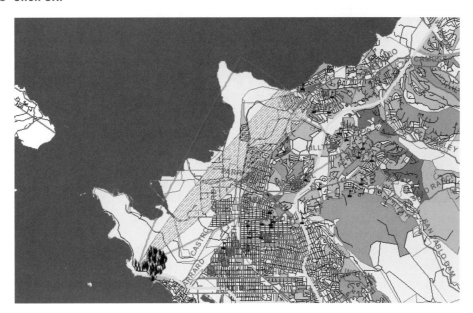

The parks and schools within a one-mile buffer outside the plume area are identified and marked.

31 Repeat this symbol marking process for the other assembly point data layers.

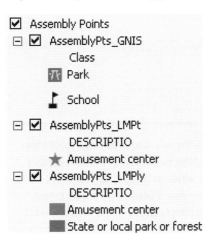

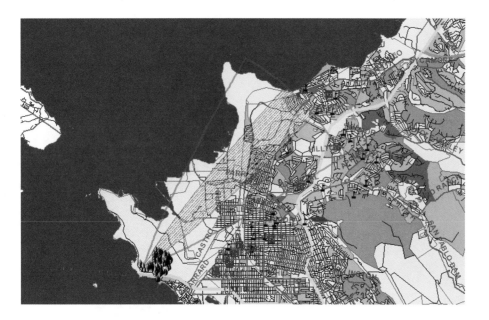

All of the possible assembly points within a one-mile buffer outside the plume area are identified and marked. By demarcating the extent of the potential plume area, residents within the impacted area can be informed via radio and automated telephone alert systems. They then can be assigned to designated assembly points based on geographic proximity and transportation routing. These sites become immediate staging areas for emergency operations personnel to assemble the impacted residents to provide emergency supplies and additional evacuation instructions. Developing tools such as this in the planning stages of a possible disaster provide the necessary tactical information homeland security and emergency operations planners need to protect people and places from the potential impacts of a catastrophic event such as this.

32 Save the ArcMap document as **FL_GISTHS_C5E6.mxd**, with FL being your initials, in **\ESRIPress \GISTHS\MYGISTHS_Work**.

YOUR TURN

An effective way to control access to an area impacted by a plume event is to place roadblocks at intersections where roads cross into the affected area. Using the select and intersect tools (as done in exercise 5.3), place a point at the intersections where primary (FCC codes A10-25) and secondary roads (FCC codes A30-39) cross into the confidence area of the plume extent. To provide even greater access control, place a point at the intersections where local streets intersect with the extent of the actual plume footprint. This ensures that no one can gain access to the area while the toxic plume event is in progress.

33 Exit ArcMap.

Summary

In this chapter you applied the suite of ArcMap tools and geospatial analysis to the *Protect* Mission Area of homeland security planning and operations. This mission area focuses on safeguarding critical infrastructure and the safety of people impacted by emergency events. Buffering, selection, and intersect tools enable identification of critical infrastructure within an area of a potential threat. Additional 3D spatial tools, such as viewshed and line of sight, optimize positioning of surveillance sites around targets facilities.

By incorporating census data into the GIS, homeland security planners and operations personnel can effectively locate and identify how many people, including those with special needs, may be affected by an emergency event. This assists in allocating necessary relief and evacuation resources prior to an event in order to ensure the safety of those affected. Spatial selection tools also help locate assembly points for evacuation outside of an affected area.

Notes

1. Executive Order 13010–Critical Infrastructure Protection. *Federal Register*, July 17, 1996. Vol. 61, No. 138. pp. 37347-37350

2. Ibid., p. 37347

3. The Clinton Administration's Policy on Critical Infrastructure Protection: Presidential Decision Directive No. 63, White Paper, May 22, 1998.

4. *GIS for Homeland Security*, ESRI Press, p. 42.

5. *The Critical Infrastructure Protection Process, Job Aid*, U.S. Fire Administration, Critical Infrastructure Protection Information Center, www.usfa.fema.gov/cipc, May 2002.

6. *What Is CIP and Why Is It Important?*, U.S. Fire Administration, Department of Homeland Security, www.usfa.dhs.gov/fireservice/subjects/emr-isac/what_is.shtm

7. Ibid., p. 5.

8. Ibid., p. 1

9. Ibid.

10. www.ni2cie.org/carver2.asp

11. DHS *National Planning Scenarios*, Version 20.1, April 2005, p. 12-1

12. Ibid., p. 12-3

13. DHS *Overview: FY 2007 Infrastructure Protection Program Final Awards*, Updated for May 10, 2007, p. 10

14. *GIS for Homeland Security*, ESRI Press, p. 42.

15. The linear extent of the buffer zone is variable and is determined by the nature of the incident, terrain, and other conditions specific to the event.

16. These settings were obtained from the U.S. Navel Observatory Astronomical Applications Department. Web site (http://aa.usno.navy.mil) for Berkeley, California, on June 13, 2007, at 4:00 p.m.

17. Since this map uses meters as its unit of measure, the Z units are recorded as meters.

18. ArcGIS Desktop Help, *Creating a line of sight*

19. *DHS National Planning Scenarios*, Version 20.1, April 2005, pp. 6-4.

20. Ibid.

21. U.S. Environmental Protection Agency, *Computer-Aided Management of Emergency Operations (CAMEO)*, http://www.epa.gov/ceppo/cameo/index.htm

22. Skelton, Gordon, All Hazards Emergency Operations Management System (ALLHAZ) /DRIS, http://www.orau.gov/DHS_RE_Summit07/Speakers_Abstracts/Skelton_1.pdf

23. U.S. Census Bureau, Cartographic Boundary Files, http://www.census.gov/geo/www/cob/bg_metadata.html

24. The city of New Haven, Connecticut, has initiated a voluntary program in which elderly residents and persons with disabilities register with the Office of Emergency Management so that they can be notified and safely evacuated and cared for in the event of an emergency. http://www.cityofnewhaven.com/EmergencyInfo/pdfs/Evacuation_Assistance.pdf

25. http://www.ready.gov/business/plan/evacplan.html

Chapter 6:

Analyzing data for homeland security planning and operations—*Respond*

DHS National Planning Scenario 10: Natural disaster—major hurricane

---

MISSION AREA: *PREVENT*
NHC/FEMA video teleconference forecast and impact assessments.

MISSION AREA: *PROTECT*
Review impact assessments for infrastructure and rapid resource needs; request remote sensing products; run path, size and intensity models; and assess search and rescue, medical, and navigation needs.

MISSION AREA: *RESPOND*
DHS Target Capability: Search and rescue
DHS Task: Res.B.4 Conduct Search and rescue.

MISSION AREA: *RECOVER*
Prepare for secondary hazards and events; assess property and structural damage from flooding and high winds; assess and recover utility service disruptions; assess public health threat from disease and hazardous materials; and assess economic repercussions of disrupted business operations.

6.1 Prepare respond scenario map 1
Add Jackson County MEDS geodatabase features to ArcMap document

6.2 Geocode missing persons
Prepare address locator
Interactively geocode street addresses
Find longitude and latitude coordinates of an address
Add missing-persons database
Batch geocode missing-persons database
Create missing-persons status maps

6.3 Prepare respond scenario map 2
Set analysis extent
Generate slope
Calculate suitable slope
Calculate land-cover class
Calculate landscape obstacles
Calculate straight line distance from towers
Reclassify straight line distance from towers
Calculate straight line distance from power lines
Reclassify straight line distance from power lines
Combine raster datasets
Reclassify combined raster dataset
Add inundated areas

6.4 Perform respond suitability analysis
Calculate suitability

6.5 Create suitability model using ModelBuilder
Create a new toolbox and model
Add data and tools to the new model
Run the suitability model

6.6 Add the U.S. National Grid
Convert data frame coordinate system
Select a grid references system
Overlay the USNG onto the map layout

Define *respond*, to include implementing immediate actions to save lives,
protect property, and meet basic human needs in the event of a major
natural disaster
Find the location and the latitude and longitude coordinates of an address
Geocode a missing-persons database
Generate a suitability map of rescue locations
Build a model to automate the geoprocessing of the suitability analyses
Use the U.S. National Grid to identify coordinate locations of missing persons

Chapter 6

Analyzing data for homeland security planning and operations: *Respond*

This chapter focuses on using GIS tools and analysis to support the *Respond* Mission Area as identified in the DHS National Preparedness Goal. DHS defines the critical components of a successful response plan as "the "ability to direct, control, and coordinate a response; manage resources; and provide emergency public information—this outcome includes direction and control through the Incident Command System (ICS), Multiagency Coordination Systems, and Public Information Systems."[1]

The National Response Plan (NRP), as required by Homeland Security Presidential Directive 5, establishes a single, comprehensive approach to domestic incident management to prevent, prepare for, respond to, and recover from terrorist attacks, major disasters, and other emergencies.[2] It uses the framework of the National Incident Management System (NIMS) which is a systematic template for incident management at all jurisdictional levels. Though the NPR is always in effect, it is actively mobilized in response to Incidents of National Significance (INS). An INS event is defined as "an actual or potential high-impact event that requires robust coordination of the Federal response in order to save lives and minimize damage, and provide the basis for long-term community and economic recovery."[3]

With that said, however, the basic premise of the NRP is that incidents are generally handled by first responders at the lowest jurisdictional level possible. State and local resources, such as police, fire, emergency medical personnel, and interstate mutual aid will provide the first line of emergency response and incident management support. When state resources and capabilities are overwhelmed, governors may request federal assistance from the military and other national level emergency advisers and agencies.[4]

GIS tools and analysis provide vital information to support Incident Command Systems at all jurisdictional levels. First responders assigned to the scene of an incident are better able to scope out the extent and impact of a catastrophic event when provided with accurate locations of geographic features.[5] When first responders are joined by supporting forces from other jurisdictions and agencies, it is critical that the geographic information referenced is generated and analyzed in a common operating picture (COP) universal to all emergency operations personnel. Such COP applications include the National Geospatial-Intelligence Agency's Palanterra and the DHS's Integrated Common Analytical Viewer (iCAV) programs that provide real-time Web-enabled environments to manage the geographic features relevant to an incident of national significance.

In the event of a major disaster, ICS must coordinate response actions that include search and rescue, medical system support, debris clearance and management, temporary emergency power, transportation infrastructure support, law enforcement assistance, and victim identification and mortuary services.[6] This chapter focuses on GIS tools and analysis to support the search-and-rescue actions during a major hurricane.

Chapter 6: Data dictionary

Layer	Type	Layer Description	Attribute	Description
JacksonCo_MEDS.gdb	File Geodatabase	Geodatabase of sample MEDS data layers for Jackson County, Mississippi		
jack_elev	Raster Dataset	NED Digital Elevation Model		
jack_lulc	Raster Dataset	National Land Cover Data 2001 (NLCD)		
JacksonCo_County	Feature Class	Jackson Co. County polygon		
JacksonCo_GeoName	Feature Class	Jackson Co. Geographic names		
JacksonCo_Hydrog_Streams	Feature Class	Jackson Co. Stream polylines		
JacksonCo_Hydrog_WatBodies	Feature Class	Jackson Co. Waterbody polygons		
JacksonCo_Powerlines	Feature Class	Jackson Co. Powerline polylines		
JacksonCo_Streets	Feature Class	Jackson Co. Street polylines		
jack_fld1	Raster Dataset	Jackson Co. Flood signature 1		
jack_fld2	Raster Dataset	Jackson Co. Flood signature 2		
jack_fld3	Raster Dataset	Jackson Co. Flood signature 3		
JacksonCo_MP.csv	Text File	Jackson Co. Missing persons database	STATUS_1	Status of missing person at first report
		(fictitious data)	STATUS_2	Status of missing person at second report

Exercise 6.1
Prepare *respond* scenario map 1

DHS National Planning Scenario 10:
> Natural disaster—major hurricane

DHS Target Capability:
> Search and rescue

DHS Task:
> Res.B.4 Conduct search and rescue

In this scenario, a category 5 hurricane hits a major metropolitan area (MMA). DHS National Planning Scenario 10: Natural disaster—major hurricane, is described as follows:

> This scenario represents a Category 5 hurricane that makes landfall at an MMA. Sustained winds are at 160 mph with a storm surge greater than 20 feet above normal. As the storm moves closer to land, massive evacuations are required. Certain low-lying escape routes are inundated by water anywhere from five hours before the eye of the hurricane reaches land.[7]

This scenario has since played out almost exactly as predicted when Hurricane Katrina struck the Gulf Coast of the United States in August 2005, just four months after the National Planning Scenarios were released. Many lessons were learned from this event, from how to better shore up levee systems along the coast and around the New Orleans metropolitan area, to more effectively coordinating response efforts across jurisdictional agencies and boundaries.

The scenario profile includes a synopsis of how the *Respond* Mission Area is to be activated in this natural disaster:

> Search-and-rescue operations: Locate, extricate, and provide on-site medical treatment to victims trapped in collapsed structures. Victims stranded in floodwater must also be located and extracted.[8]

GIS tools and analysis played a critical role in locating missing persons during the search-and-rescue response to Hurricane Katrina. More than 10,000 calls came in to the Mississippi Emergency Management Agency from friends and relatives in communities along the coast. These reports included information about who the missing persons were, where and when they were last seen, and what special needs they had that would require additional assistance. This information was then added to a Web-enabled database that was managed and used by various government and relief agencies to locate and assist these people in need.[9]

In the event of a hurricane, where much of the low-lying land is inundated by floodwaters, the conventional locator designation—street address—is no longer a viable indicator of location. Much of the street network may be under water and not accessible to ground-based search-and-rescue personnel or visible to airborne operations. During the Katrina search-and-rescue operation, street addresses were converted to geographic x,y coordinates and plotted on a map to show the locations and status of missing persons. Additional GIS tools were employed to cluster these locations and embed them in the U.S. National Grid coordinate system for compliance with Coast Guard search-and-rescue operations.

In this chapter, you will use a geodatabase of geographic features along the Gulf Coast of Mississippi in Jackson County to geocode reported street addresses of the last known location of missing persons. In addition, you will perform a suitability analysis to locate operative landing sites for search-and-rescue helicopters close to the reported missing-persons locations.

Add Jackson County MEDS geodatabase features to ArcMap document

1   Open a new ArcMap document.

2   Open ArcCatalog.

3   Navigate to the **\ESRIPress\GISTHS\GISTHS_C6** folder and review the contents of the JacksonCo_
    MEDS geodatabase.

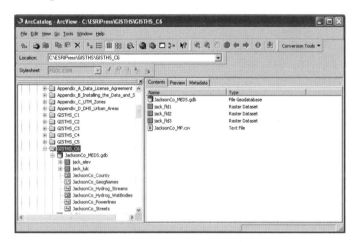

4   Drag the following features from ArcCatalog to add them to the new ArcMap document:
        JacksonCo_County
        JacksonCo_Hydrog_Streams
        JacksonCo_Hydrog_WatBodies
        JacksonCo_Streets

5   Change the symbology of each of these features and order them as follows:

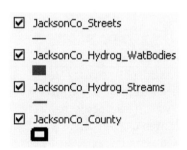

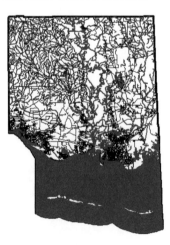

The map of Jackson County, Mississippi is added to the ArcMap document.

6   Save the ArcMap document as **FL_GISTHS_C6E1.mxd**, with FL being your initials, in **\ESRIPress
    \GISTHS\MYGISTHS_Work**.

Exercise 6.2
Geocode missing persons

Prepare address locator

During a major hurricane such as Katrina, streets may be inundated by storm surges and floodwaters that render the street address useless as an indicator of location. Geocoding each last known location of a missing person places a point on the map at its x,y coordinates.

Geocoding a database of addresses begins by creating an address locator in ArcCatalog. The address locator is generated from a street feature class or shapefile that contains a range of possible addresses on the left and right sides of the street. Each address is searched along the street and positioned on either the right or left, depending on if the address is odd or even.

1   In ArcCatalog, navigate to **\ESRIPress\GISTHS\MYGISTHS_Work**.

2   From the File menu, select New Address Locator.

3   Scroll down and select the Address Locator Style US Streets.

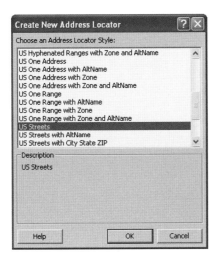

4   Click OK.

5   Name the address locator **FL_GISTHS_C6_JackCo**, with FL being your initials.

6   Click the browse button to select Reference data.

7   Navigate to **ESRIPress\GISTHS_C6** and expand the contents of the JacksonCo_MEDS geodatabase.

8   Choose JacksonCo_Streets and click Add.

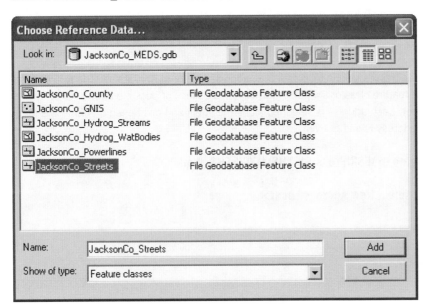

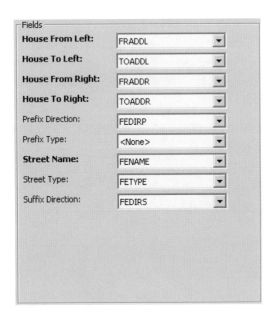

The to and from address, and street name fields are loaded into the Address Locator form.

9   Set the Side offset option to 25 feet.

Since addresses are actually located on either side of the street, rather than along the street centerline, this offsets the placement of the point 25 feet on either side.

10   Click to check that the output fields include the x,y coordinates.

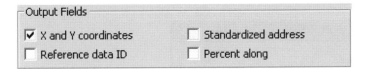

11   Click OK.

12   Scroll down in ArcCatalog see this address locator listed in **\ESRIPress\GISTHS\MYGISTHS_Work**.

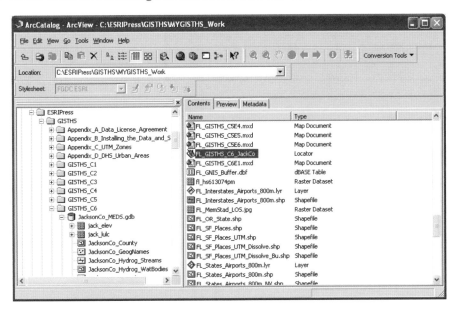

Interactively geocode street addresses

During the first few hours and days of Hurricane Katrina, before a comprehensive missing-persons database could be compiled, the Mississippi Emergency Management Agency (MEMA), U.S. Coast Guard, and GIS volunteers found several immediate uses for GIS. One of the most high-impact uses was to find the physical location on a map of an individual person in a life-threatening situation and identify the longitude and latitude of that site. This enabled immediate and direct help to that specific location, be it by ground or air rescue.

EXERCISES

ArcMap is able to interactively geocode individual addresses so that they can be located and labeled as they come into an emergency operations dispatch. While this does not produce a final systematic data management process or report, this simple location function was of critical importance to early first responders as they began their Hurricane Katrina search-and-rescue operation.

1   In ArcMap, on the Tools toolbar, click the Find button.

2   Click the Addresses tab.

3   Navigate to the address locator, FL_GISTHS_C6_JackCo, that you created earlier.

4   Enter the address **3388 Jamaica Dr** into the Street or Intersection field.

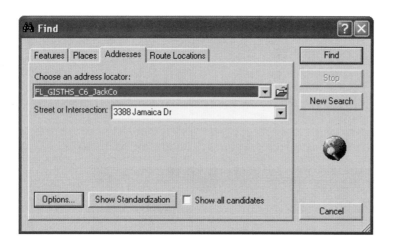

5   Click Find.

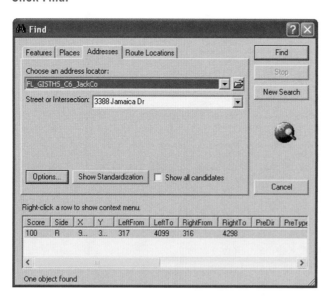

One object, or street address, is found in the address locator.

6   Shift the Find box away from the map display.

7   Right-click the address record and select Zoom to.

The map zooms to and flashes a large green symbol marker at this location.

8   Right-click the address record again and select Add Labeled Point.

A symbol marker and address label are added to the map display. This is a quick and easy address locator function that enables an early first responder to immediately direct search-and-rescue resources as needed to this location.

9   Close the Find box.

Find longitude and latitude coordinates of an address

In addition, the longitude and latitude coordinates can also be reported to assist first responders who may be using GPS systems using this geographic coordinate system.

1   Set the Display Units in the Data Frame Properties box to Degrees Minutes and Seconds.

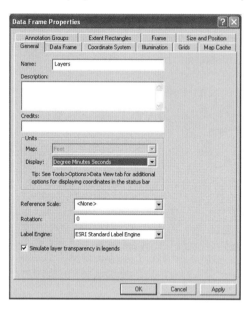

2   Place the cursor over the point and read the longitude and latitude of the point from the lower right corner of the map display.

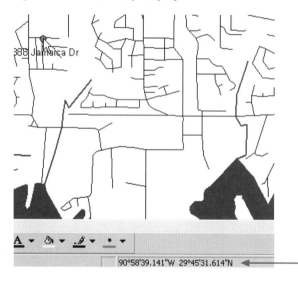

This simple address and coordinate locator function was critically important to first responders at the early stages of the Hurricane Katrina search-and-rescue operation. As the event unfolded and operating systems were put in place, more systematic and comprehensive information processing was performed to support search-and-rescue operations.

YOUR TURN

Use the Find function to interactively geocode, label, and note the latitude and longitude of additional addresses at the following locations: 1693 Dana Ct, 3310 Dijon Ave, and 200 Shiloh Cir.

3   From the Edit Menu, select Select All Elements.

4   Hit the delete key to remove the markers and labels from the map display.

5   Click the Refresh button 🔄 at the lower left corner of the map display if the labels don't disappear when removed.

6   Save the ArcMap document as **FL_GISTHS_C6E2.mxd**, with FL being your initials, in **\ESRIPress \GISTHS\MYGISTHS_Work**.

Add missing-persons database

After the early stages of response to Hurricane Katrina, a Web-enabled database was put in place to maintain a list of missing persons reported to emergency dispatch. These records could then be systematically batch geocoded to find the locations of all reported missing persons. Once these locations were added to the map, search-and-rescue personnel not only had specific location points for each person but an overall picture of where their resources would be best allocated throughout the region.

1   Add JacksonCo_MP.csv from \ESRIPress\GISTHS\GISTHS_C6 to the map document.

2   Open JacksonCo_MP.csv to review the contents of the database.

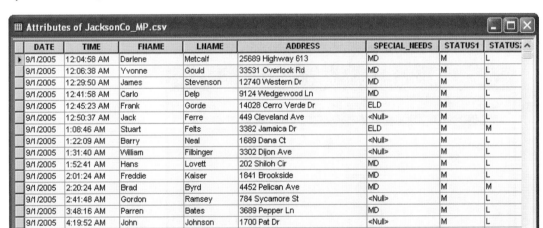

The missing-persons database contains a record for each person reported missing in Jackson County during an event such as Katrina.[10] In addition to the date, time, name, and address of the last known location of the individual, additional fields in the database contain information regarding the special needs and ongoing search-and-rescue status of the missing person. Additional relevant data fields may include age, telephone number, and family contact person. This database is simplified for use in this tutorial.

3   Close the Attributes of JacksonCo_MP.csv table.

Batch geocode missing-persons database

The missing-persons database enables batch geocoding by processing a full set of addresses all at once. The records in the database may be compiled from an ongoing list of calls coming into a 911 dispatch or emergency operations center.

1   Right click JacksonCo_MP.csv and select Geocode Addresses.

2   If not already added, click Add and navigate to **\ESRIPress\GISTHS\MYGISTHS_Work** to select the FL_GISTHS_C6_JackCo address locator created earlier.

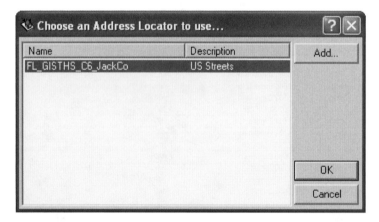

3   Click OK.

4   Output the shapefile of locations as **FL_Geocoding_Result_MP.shp**, with FL being your initials, in **\ESRIPress\GISTHS\MYGISTHS_Work**.

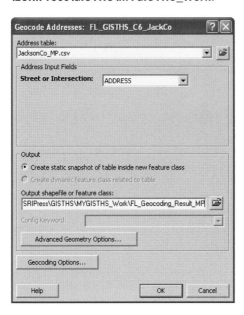

5   Click OK.

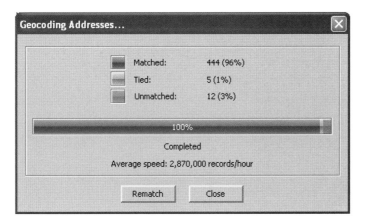

The resulting box shows 444 addresses (96 percent were matched to the address locator streets), five (1 percent) tied, and 12 (3 percent) unmatched. Possible reasons for unmatched addresses are data entry errors that include incorrect house numbers or misspelled street names. Unmatched addresses can be interactively matched by manually comparing each address to a set of candidates selected by the system, and then adding them to the geocoded results. Since 97 percent of this database matched to the address locator, you will accept these results.

6   Click Close.

The geocode results are added to the map display as a set of points.

7   Zoom to the extent of the geocoding results data layer to view the full set of geocoded address points.

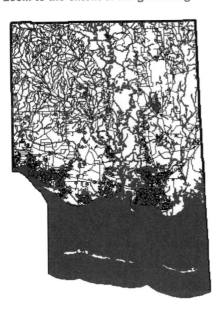

8   Zoom into an area of dense points to see how they are positioned along the street features.

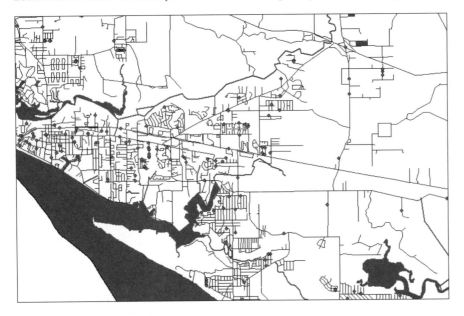

Emergency personnel now know the last known locations of missing persons and can use this map to plan search-and-rescue operations.

EXERCISES

Create missing-persons status maps

The missing-persons database includes three status fields that indicate if the person is missing (M) or located (L). STATUS1 is the initial missing person designation; STATUS2 and STATUS3 are instances later in time when emergency personnel updated the database to review status changes. Status maps can be generated at designated time intervals to track the ongoing status changes as search-and-rescue operations track and recover missing persons throughout the response phase of an emergency event.

1   Right-click Geocoding Result: FL_Geocoding_Result_MP and select Properties.

2   Change the symbology and label of the STATUS1 field to a red pushpin to mark the location of each missing person.

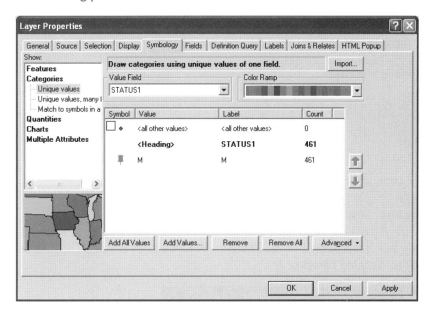

3   Zoom to an area to see more closely the distribution of missing persons.

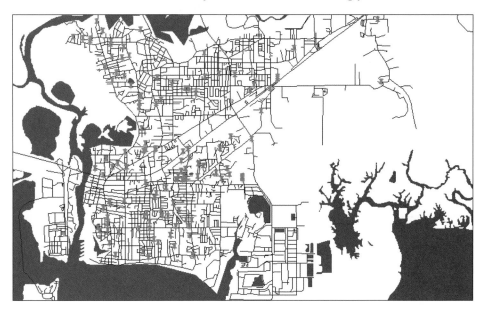

4   Right-click Geocoding_Result:FL_Geocoding_Result_MP and select Copy.

5   From the Edit menu, select Paste.

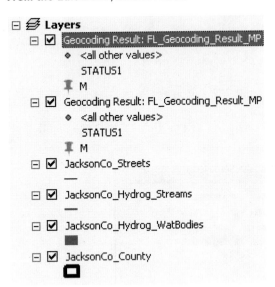

An identical copy of the geocoding result layer is added to the table of contents.

6   Rename this new data layer **Geocode: STATUS2**.

7   Change the symbology and labels of the STATUS2 field to the same symbol as STATUS1, adding a green pushpin marker for Located Persons.

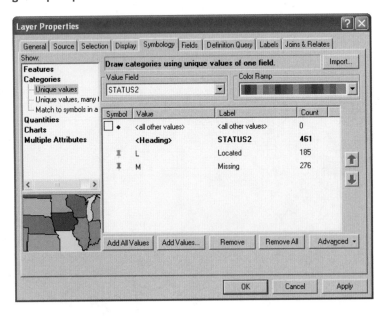

STATUS1

STATUS2

This updated map shows persons located and still missing at this later time. Toggling the STATUS2 data layer on and off shows this change effectively as red markers change to green as missing persons are located.

YOUR TURN

There is also a field in the missing-persons database for special needs. Select only the STATUS2 missing persons and create a new data layer to map the special needs of this population.

8   Save the map document when complete.

EXERCISES

Exercise 6.3
Prepare respond scenario map 2

Now that the locations of the people reported missing during a major hurricane have been located on the map, emergency personnel need to devise a search-and-rescue plan to get to those locations, find these people, and return them to safety.

Since some locations may not be accessible by ground-based rescue vehicles, sites suitable for helicopter landing zones need to be located. This analysis was performed by the U.S. Army Space and Missile Defense Command/U.S. Army Forces Strategic Command (SMDC/ARSTRAT) Measurement and Signature Intelligence/ Advanced Geospatial Intelligence (MASINT/AGI) Node in response to Hurricane Katrina, and was highlighted in an article in ArcNews in 2006.[11] This exercise is a simplified adaptation of the steps and procedures SMDC/ARSTRAT used to perform this analysis.

The criteria analyzed to locate suitable helicopter landing zones included slope of the terrain, land cover classes, and landscape obstacles such as towers and power lines. In addition, temporally updated maps of terrain altered by flooding were generated from high resolution multispectral satellite imagery that was analyzed and classified to show areas of inundation.[12]

Set analysis extent

1   From the Main menu, click Tools, Extensions.

2   Click the box beside Spatial Analyst and click Close.

4   From the Main menu, click View, Toolbars, and select Spatial Analyst.

5   Dock this menu bar in a spot at the top or side of the map display.

6   From the Spatial Analyst toolbar, select Options.

7   Click the Extent tab.

8   Select Same as Layer "JacksonCo_County" as the Analysis extent.

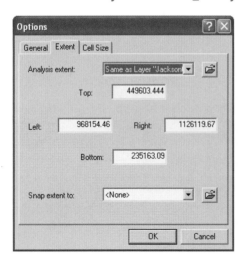

9   Click OK.

Generate slope

Slope is defined as the incline from a horizontal or vertical position.[13] The Spatial Analyst extension to ArcMap includes a slope function that identifies the steepest downhill slope for a location on a surface.[14] Slope can be reported as either degrees or percentage. When the slope angle equals 45 degrees, the rise is equal to the run. Expressed as a percentage, the slope of this angle is 100 percent. As the slope approaches vertical (90 degrees), the percentage slope approaches infinity.[15] Helicopters require a slope of less than 15 percent to land safely.

1   Turn off all data layers except JacksonCo_County and zoom to its extent.

2   Add the jack_elev raster dataset from the JacksonCo_MEDS geodatabase to the ArcMap document.

The digital elevation model showing the elevation of Jackson County terrain is added to the map display.

3   From the Spatial Analyst toolbar, select Surface Analysis > Slope.

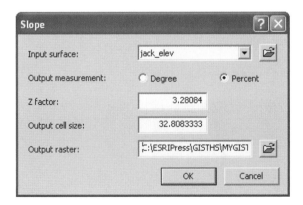

The jack_elev raster dataset is already selected as the input surface.

4   Click Percent as the output measurement.

5   Enter a Z factor of 3.28084. This raster dataset uses feet as the x,y units of measure and meters as the z unit of measure. Since there are 3.28084 feet in a meter, this Z factor takes this conversion into account when calculating slope.

6   Accept the output cell size, and save the output raster as **FL_jack_slope**, with FL being your initials, in **\ESRIPress\GISTHS\MYGISTHS_Work**.

7   Click OK.

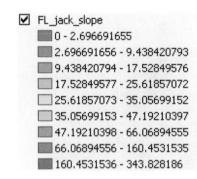

The slope map is added to the map display. The slope percentage shows the areas with the lowest slope in green ramping to the highest in red. Since the most suitable slope percentage for a helicopter landing zone is less than 15 percent, you can isolate only those areas that meet that criteria by using the Raster Calculator.

Calculate suitable slope

1   From the Spatial Analyst toolbar, select Raster Calculator.

2   Double-click FL_jack_slope to enter it into the expression box.

3   Click the Less Than button $<$ to enter it into the expression box.

4   Enter **15**.

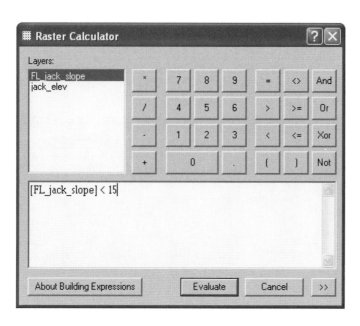

5   Click Evaluate.

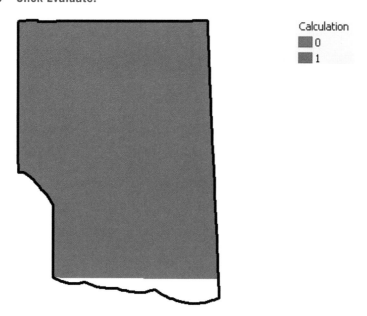

A new raster dataset is added to the map display showing the areas that are less that 15 percent slope with a value of 1, and the areas at or above 15 percent slope with a value of 0. This is a temporary calculation dataset which you will use for further calculations, so you will save it as a permanent data layer.

6   Right-click Calculation and select Data > Make Permanent.

7   Save the output raster as **FL_jack_sl15**, with FL being your initials, in **\ESRIPress\GISTHS \MYGISTHS_Work**.

8   Right-click Calculation again and select Properties.

9   Select the General tab and change the name to **FL_jack_sl15**.

EXERCISES

10 Remove jack_elev and FL_jack_slope from the Table of Contents.

11 Save the ArcMap document as **FL_GISTHS_C6E3.mxd**, with FL being your initials, in **\ESRIPress \GISTHS\MYGISTHS_Work**.

You now have a raster dataset that shows all of the sites potentially suitable for a helicopter landing zone based solely on the slope of the terrain. All other areas will be excluded from consideration when determining suitable areas for this use. Now you can analyze additional criteria to refine the suitability analysis for this study.

Calculate land-cover class

The National Land Cover Database of 2001 classifies 16 categories of land cover. Areas that are not suitable for helicopter landing zones include urban/built-up areas, agricultural areas, forested or heavily vegetated areas, wetlands, and water. Areas suitable for these zones include rural or low to medium developed areas, barren land, scrub and grassland areas, and pastures. The raster calculator can be used to select only those areas that are suitable for helicopter landing zones.

1 Add the jack_lulc raster dataset from the JacksonCo_MEDS geodatabase to the ArcMap document.

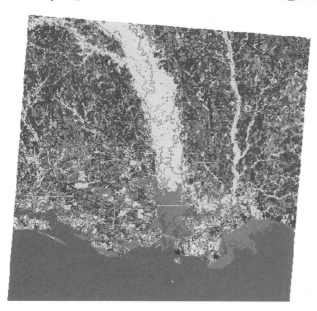

The raster dataset showing the land classes in Jackson County is added to the map display.

2 Use the Raster Calculator to select the following land cover classes:

Note: Use "or" for the pipe separator between classes.

- **21 Developed, Open Space, or**
- **22 Developed, Low Intensity, or**
- **31 Barren Land (Rock/Sand/Clay), or**
- **51 Dwarf Scrub, or**
- **52 Shrub/Scrub, or**

- • **71 Grassland/Herbaceous, or**
- • **72 Sedge/Herbaceous, or**
- • **73 Lichens, or**
- • **74 Moss, or**
- • **81 Pasture/Hay**

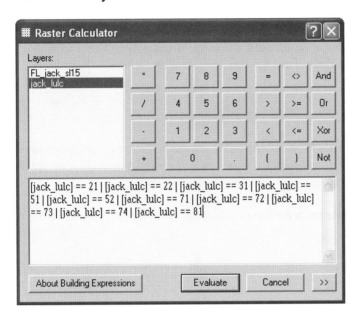

3   **Click Evaluate.**

A new raster dataset is added to the map display showing, in this display, the suitable land classes in violet with a value of 1, and the unsuitable land classes in brown with a value of 0.

Note: Colors are randomly assigned and may differ in your map display.

4 Make this a permanent raster dataset named **FL_jack_slc**, with FL being your initials, and rename it in the Table of Contents.

5 Remove jack_lulc from the Table of Contents.

6 Save the map document when compete.

You now have a raster dataset that shows all of the sites potentially suitable for a helicopter landing zone based solely on the class of land cover. All other areas will be excluded from consideration when determining suitable areas for this use.

Calculate landscape obstacles

Objects on the landscape that pose obstacles to helicopters include transmission and observation towers, and power lines and pipelines. Placing a 50-foot buffer around these features ensures that they will be avoided when locating suitable helicopter landing zones.

1 Add the JacksonCo_GeogNames data layer from the JacksonCo_MEDS geodatabase to the ArcMap document.

2 From the Selection menu, Select by Attributes from JacksonCo_ GeogNames those features where "Class" = 'Tower'.

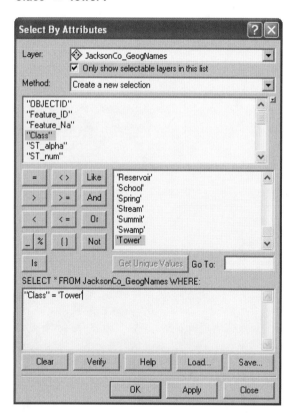

Calculate straight line distance from towers

1 From the Spatial Analyst toolbar, select Distance > Straight Line.

2   Select JacksonCo_GeogNames as Distance to: and enter **91.2011468** as the Output cell size.

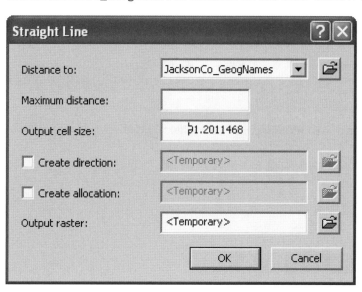

The slope raster dataset has a cell size of 32.808333, or 10 meters. The land cover raster dataset has a cell size of 91.2011468, or approximately 30 meters. When combining raster datasets for analysis, the resampled cell size will be at the largest cell size so as not to impose a false resolution and accuracy on the larger cell sizes than originally acquired. Since the land cover cell size is the larger of the two, it is the cell size to choose for this raster dataset.

3   Click OK.

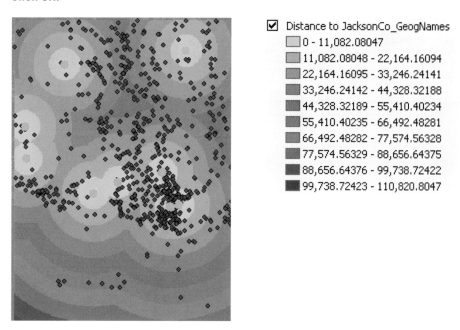

A new raster dataset is added to the map display showing a series of buffer rings around each selected tower feature. Each ring extends approximately 11,000 feet. In order to place a 50-foot buffer around each of these towers, the dataset needs to be reclassified.

Reclassify straight line distance from towers

1   From the Spatial Analyst toolbar, select Reclassify.

2   Select Distance to JacksonCo_GeogNames as the input raster.

3   Edit the first Old values field to 0-50 and New values field to 0.

4   Edit the next Old values field to 51-110820.8047 and New values field to 1.

5   Hold down the shift key and click to select all remaining rows except the last NoData row and click Delete Entries.

6   Save to output raster as **FL_jack_tower**, with FL being your initials, in **\ESRIPress\GISTHS \MYGISTHS_Work**.

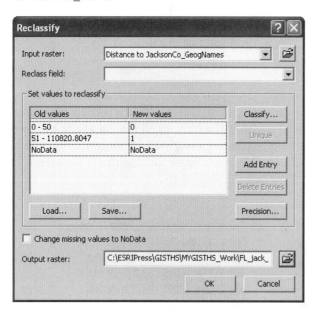

7   Click OK.

A new raster dataset is added to the map display showing the suitable locations outside of the 50-foot buffer with a value of 1, and the unsuitable locations within the 50-foot buffer with a value of 0.

8   Zoom close into one of the selected JacksonCo_GeogNames features.

9   Turn off the JacksonCo_GeogNames data layer to reveal the location of this tower's 50-foot buffer.

You now have a raster dataset that shows all of the sites potentially suitable for a helicopter landing zone based solely on the location of this landscape obstacle. All locations within a 50-foot buffer around a tower will be excluded from consideration when determining suitable areas for this use.

10   Remove JacksonCo_GeogNames and Distance to JacksonCo_GeogNames from the Table of Contents.

11   Save the map document when compete.

Calculate straight line distance from power lines

1   Add JacksonCo_Powerlines from the JacksonCo_MEDS geodatabase to the ArcMap document.

2   Zoom to the extent of the JacksonCo_County data layer.

3   From the Spatial Analyst toolbar, select Distance > Straight Line.

4   Select JacksonCo_Powerlines as Distance to: and enter **91.20114679** as the Output cell size.

5   Click OK.

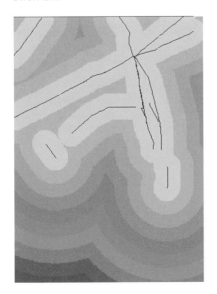

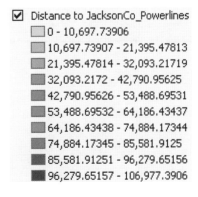

☑ Distance to JacksonCo_Powerlines
   ☐ 0 - 10,697.73906
   ☐ 10,697.73907 - 21,395.47813
   ☐ 21,395.47814 - 32,093.21719
   ☐ 32,093.2172 - 42,790.95625
   ☐ 42,790.95626 - 53,488.69531
   ☐ 53,488.69532 - 64,186.43437
   ☐ 64,186.43438 - 74,884.17344
   ☐ 74,884.17345 - 85,581.9125
   ☐ 85,581.91251 - 96,279.65156
   ☐ 96,279.65157 - 106,977.3906

A new raster dataset is added to the map display showing a series of buffer rings around each power line. Each ring is approximately 10,000 feet. In order to place a 50-foot buffer around each of these power lines, the dataset needs to be reclassified.

Reclassify straight line distance from power lines

1   From the Spatial Analyst toolbar, select Reclassify.

2   Select Distance to JacksonCo_Powerlines as the input raster.

3   Edit the first Old values field to 0 - 50 and New values field to 0.

4   Edit the next Old values field to 51-106977.390625 and New values field to 1.

5   Hold down the shift key and click to select all remaining rows except the last NoData row and click Delete Entries.

6   Save to output raster as **FL_jack_power**, with FL being your initials, in **\ESRIPress\GISTHS \MYGISTHS_Work**.

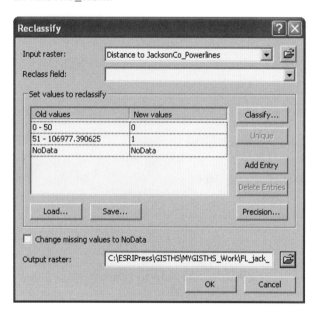

7   Click OK.

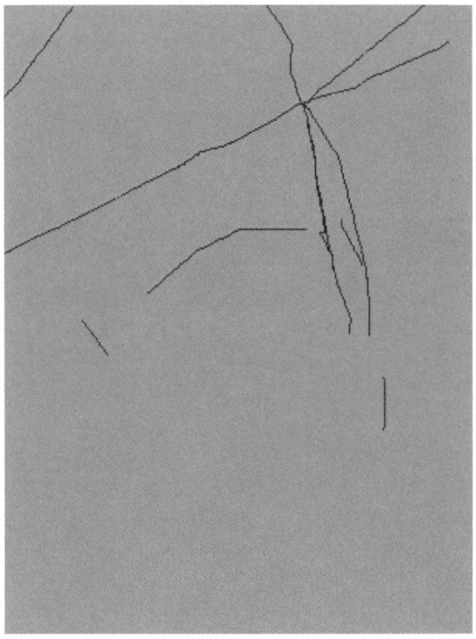

You now have a raster dataset that shows all of the sites potentially suitable for a helicopter landing zone based solely on the location of power lines. All locations within a 50-foot buffer around the tower and power line obstacles will be excluded from consideration when determining suitable areas for this use. Combining this raster dataset with the towers raster dataset will result in a new raster dataset that shows all of the sites outside of all of these landscape obstacles.

8 **Remove JacksonCo_Powerlines and Distance to JacksonCo_Powerlines from the Table of Contents.**

9 **Save the map document when compete.**

Combine raster datasets

1 **From the Spatial Analyst toolbar, select Raster Calculator.**

2 **Enter the expression [jack_tower] + [jack_power].**

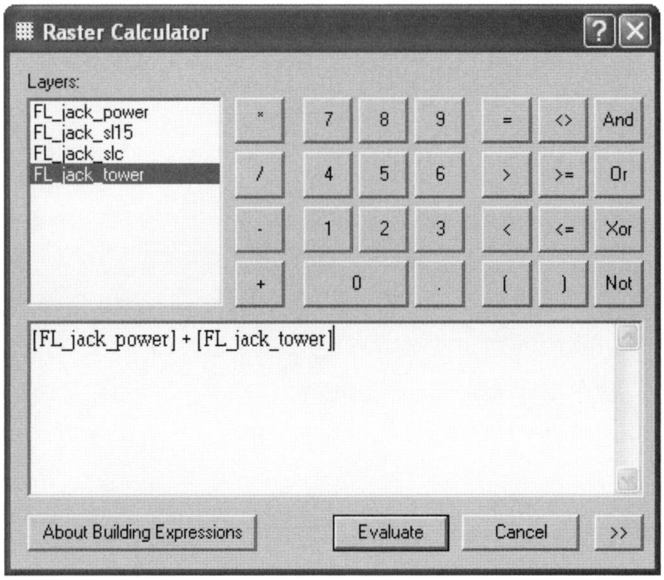

3   Click Evaluate.

Calculation
■ 1
■ 2

A new raster dataset is added to the map display showing the combined locations of towers and power lines. Because there are no concurrent locations where both towers and power lines intersect, there are no cells with a value of zero. The cells with a value of 1 are those locations where either a tower or power line is located. The cells with a value of 2 are those locations where neither of these landscape features is located, and are therefore suitable for helicopter landing sites. These values need to be reclassified for consistent analysis along with the other suitability data layers.

Reclassify combined raster dataset

1   From the Spatial Analyst toolbar, select Reclassify.

2   Select Calculation as the input raster.

3   Edit the first New values field to 0.

4   Edit the next New values field to 1.

5 Save to output raster as **FL_jack_obs**, with FL being your initials, in **\ESRIPress\GISTHS \MYGISTHS_Work**.

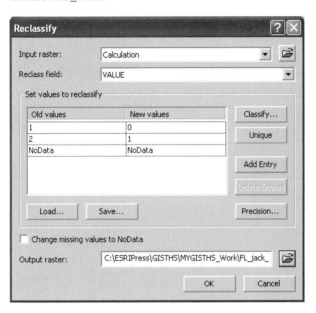

6 Click OK.

7 Remove Calculation, FL_jack_power, and FL_jack_tower from the Table of Contents.

In addition to the geodatabase MEDS and geocoded data layers, only the following raster datasets remain in the table of contents:

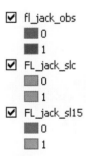

☑ fl_jack_obs
  ■ 0
  ■ 1
☑ FL_jack_slc
  ■ 0
  ■ 1
☑ FL_jack_sl15
  ■ 0
  ■ 1

8 Save the map document when compete.

Add inundated areas

In the wake of Hurricane Katrina, GIS analysts at the U.S. Army Space and Missile Defense Command/U.S. Army Forces Strategic Command (SMDC/ARSTRAT) Measurement and Signature Intelligence/Advanced Geospatial Intelligence (MASINT/AGI) Node updated maps of terrain altered by flooding to show areas of inundation. For this exercise, fictitious inundation area signatures for three time periods prepared from Landsat imagery of Jackson County, Mississippi, are used to replicate the analysis.

EXERCISES

1   Navigate to **C:\ESRIPress\GISTHS\GISTHS_C6** and add jack_fld1 to the ArcMap Document.

2   Change the symbology so the areas of inundation and standing water (value 0) are blue and all other areas (value 1) are No Color.

    ☑ jack_fld1
      ■ 0
      □ 1

3   Turn on the JacksonCo_Streets and Geocoding: STATUS2 data layers.

4   Drag the jack_fld1 raster layer below these data layers so you can see where the streets and missing persons are located.

5   Zoom to an area along the coast where there is a concentration of missing persons.

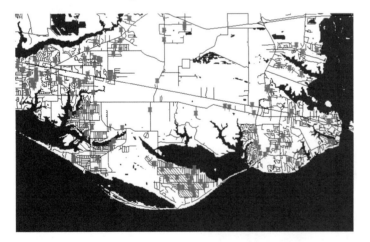

Note that there are areas along the coast and inland waterways that are inundated because of the hurricane storm surge and subsequent flooding. In order for search-and-rescue personnel to gain access to some of these areas, they must know where it is safe to land. By combining the suitability data layers built in this exercise, you can determine those locations where all of the suitable conditions exist, and hence, are safe for helicopter landing zones.

6   Save the map document when complete.

Exercise 6.4
Perform respond suitability analysis

Each of the suitability data layers prepared in the earlier exercise shows locations with conditions suitable for helicopter landing zones. Each raster cell in each raster dataset is identified by either 0 or 1. Zero (0) means that the cell does not meet the criteria and, conversely, one (1) means that the cell does meet the criteria. For a single location to be suitable for a helicopter landing zone, it must contain ALL of the following conditions: slope less than 15 percent; land uses outside of highly developed and agricultural areas; at least 50 feet from landscape obstacles; and not be inundated by storm surge or standing water. By combining these raster datasets, each cell is a sum of the four values. Only the cells with a combined value of four (4) meet all of the criteria and are suitable for helicopter landing zones.

Calculate suitability

1   From the Spatial Analyst toolbar, select Raster Calculator.

2   Enter the expression [FL_jack_obs] + [FL_jack_sl15] + [FL_jack_slc] + [jack_fld1].

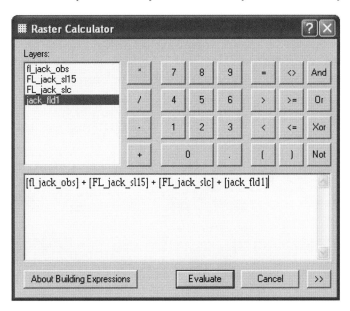

3   Click Evaluate.

4 Zoom to the extent of the new Calculation layer.

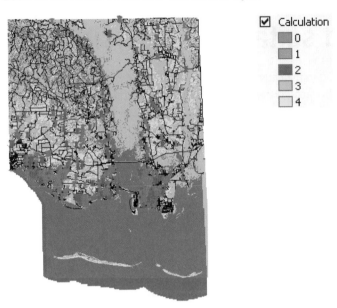

You now have a raster dataset that shows all of the sites potentially suitable for helicopter landing zones based on the combined criteria of slope, land cover, obstacles, and inundation. Only the cells with a combined value of four (4) meet all of the criteria and are suitable for helicopter landing zones.

5 Reclassify the Calculation raster dataset and save it as **FL_jack_hlz1**, with FL being your initials, in **\ESRIPress\GISTHS\MYGISTHS_Work**.

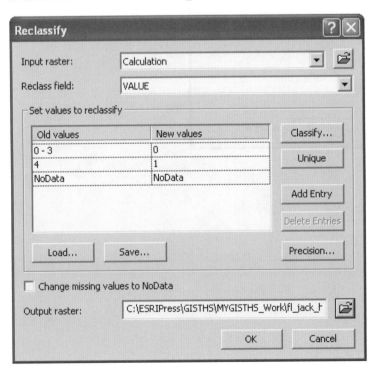

6 Change the symbology of the FL_jack_hlz1 raster dataset so that zero is gray and one is No Color.

☑ jack_hlz1
  ▨ 0
  ☐ 1

7 Remove the Calculation layer, turn off all other raster data layers, and turn on JacksonCo_Hydrog_ WatBodies. If needed, drag FL_jack_hlz1 below the JacksonCo_Hydrog_WatBodies layer.

8 Zoom to an area where there is a concentration of missing persons.

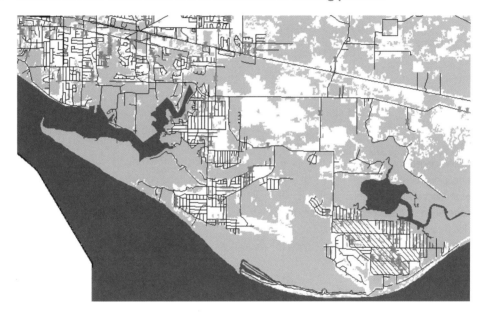

On this map, only the white areas are suitable landing zones. They are below a 15 percent slope, are outside of highly developed and agricultural lands, are beyond a 50-foot buffer from towers and power lines, and are not inundated by storm surges or standing water. Search-and-rescue personnel can use this map to determine the optimal location to land in proximity to the geocoded locations of reported missing persons.

YOUR TURN

Perform this suitability analysis again, this time using the jack_fld2 raster dataset to see how the updated areas of inundation change the location of suitable helicopter landing zones.

9 Save the ArcMap document as **FL_GISTHS_C6E4.mxd**, with FL being your initials, in **\ESRIPress \GISTHS\MYGISTHS_Work**.

Exercise 6.5
Create suitability model using ModelBuilder

ArcMap includes a set of tools to create models using the ArcToolbox and ModelBuilder applications. Creating a model automates a sequence of operations that results in the creation of new geographic information. As the inundation conditions along the coast changed during the Hurricane Katrina storm surge, new suitability analyses were performed to reevaluate the changing suitability of helicopter landing zones. In this exercise, you will build a simple model to automate the creation of a new raster dataset showing suitable helicopter landing zones based on the fixed variables of slope, land-cover, and landscape obstacles, and the changing variable of flood inundation.

Create a new toolbox and model

1   Open ArcToolbox.

2   Right-click anywhere in the white area of ArcToolbox and select New Toolbox.

A new toolbox appears in the ArcToolbox Favorites list.

3   Rename this tool **HLZ_Model**.

4   Right-click HLZ_Model and select New > Model.

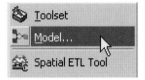

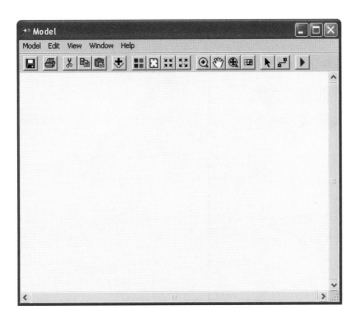

An empty Model window appears.

Add data and tools to the new model

1   From the Table of Contents, drag and drop the following raster datasets into the Model window:

-    **FL_jack_obs (with FL being your initials)**
-    **FL_jack_sl15**
-    **FL_jack_slc**

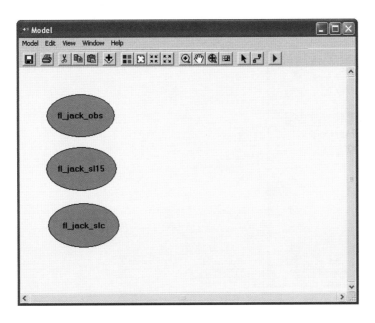

Each input raster dataset appears as a blue oval model element in the Model window. These are the fixed conditions of slope, land cover, and obstacles that do not change in the analysis.

2   Scroll in ArcToolbox to the Spatial Analyst tools and expand the Math set of tools.

3   Drag and drop the Plus tool into the Model window.

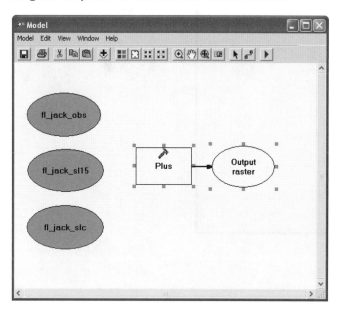

This tool appears as a white rectangle model element with an arrow extending to a white oval model element labeled Output raster.

4   Click the Add Connection button [icon] from the ModelBuilder toolbar and drag a connection line from FL_jack_obs to the Plus box.

5   Select Input raster or constant value 1 (Parameter) and click OK.

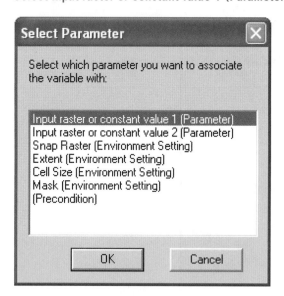

6   Click the Add Connection button from the ModelBuilder toolbar and drag a connection line from FL_jack_sl15 to the Plus box.

7   Select Input raster or constant value 2 (Parameter) and click OK.

When they are connected, the Plus and Output raster boxes fill with color to show that the model now has enough data to run.

The Plus tool enables the calculation of only two raster datasets at a time. To combine more than two raster datasets, the Plus tool needs to be repeated and the resulting output raster datasets are then combined. In this case, however, the third raster dataset is a fixed dataset, fl_jack_slc, while the fourth raster dataset is a variable one based on the latest inundation raster dataset is to be analyzed. ModelBuilder enables the creation of a parameter to select the variable input raster dataset to use in the model as needed.

8   Drag another Plus tool into the Model window, and connect the FL_jack_slc model element.

9   Select Input raster or constant value 1 (Parameter) and click OK.

10   Right-click the Plus (2) box and select Make Variable > From Parameter > Input raster or constant value 2.

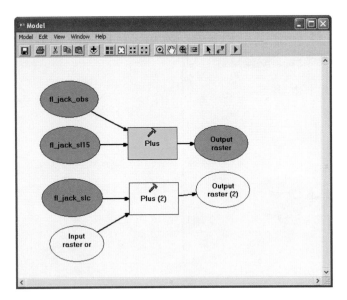

A new Input raster oval appears with a connection to the Plus box.

11 Drag the new oval below the other model elements to better position it in the Model window, and enlarge the window if necessary.

12 Right-click the new Input raster oval and select Model parameter.

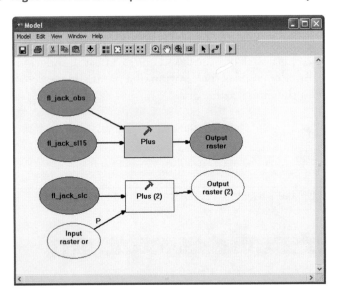

A **P** is placed along the connection line to indicate that this input raster is to be determined by a parameter. This will be the latest updated inundation raster dataset which will be selected for input each time the model is run.

So far, your model will run two separate raster calculations, each resulting in an output raster dataset. These two raster datasets are to be combined to arrive at the final raster dataset that calculates the new locations of suitable helicopter landing zones based on the latest inundation information entered into the model. The final raster dataset is also determined by a parameter so it can be individually homeled for each execution of the model.

13   Drag another Plus tool into the Model window.

14   Enlarge the Model window and place the Plus box to the right of the model elements.

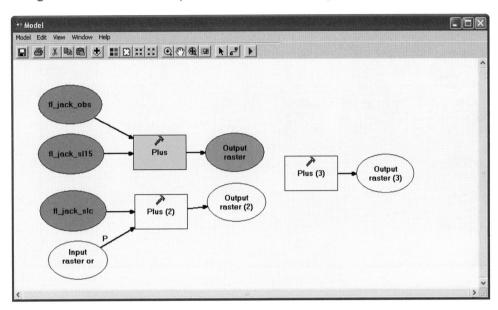

15   Connect both Output raster ovals to the new Plus (3) box, selecting the output parameters as I and 2 accordingly.

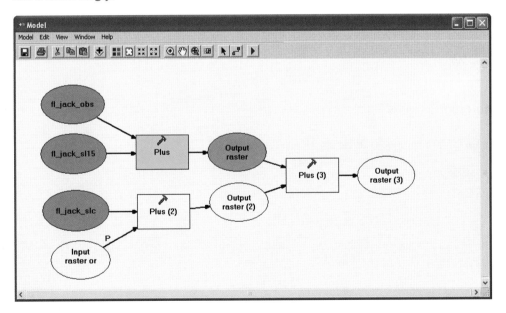

EXERCISES

16 Right-click the Output raster (3) oval and select Model parameter.

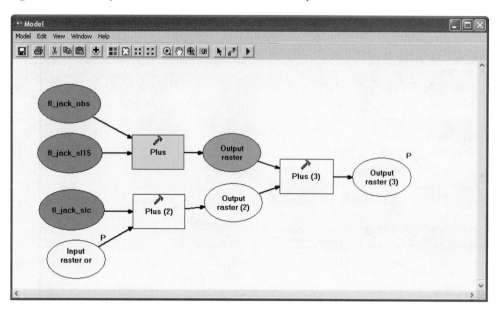

A **P** is placed above the Output raster (3) oval to indicate that this output raster is to be determined by a parameter.

The HLZ_Model is now complete.

17 Click the Save button 🖫 to save the HLZ_Model.

18 Close the Model window.

Run the suitability model

1 From the ArcToolbox, double-click the Model inside the HLZ_Model toolbox.

2 Click the browse button for the input raster, and navigate to **\ESRIPress\GISTHS\GISTHS_C6** and select **jack_fld2** as the input raster.

3   Click the browse button and navigate to **\ESRIPress\GISTHS\MYGISTHS_Work** and save the output raster as **FL_jack_hlz2**, with FL being your initials.

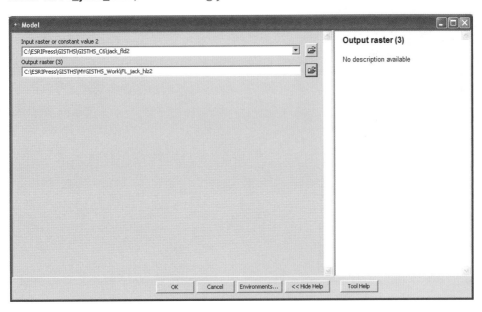

4   Click OK.

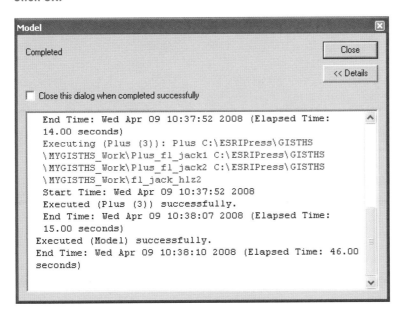

The model systematically executes through each model element and reports when it is successfully completed.

5   Click Close.

The new raster dataset is added to the map display.

6   Reclassify this new raster and compare it to the suitability raster dataset created in exercise 6.4 in the "Your turn" section. Though the colors may different, the extent of the suitable areas should be identical.

YOUR TURN

Run this HLZ_Model again using the jack_fld3 raster dataset found in the \ESRIPress\GISTHS\ GISTHS_C6 folder. Each iteration of this model using the jack_fld1, jack_fld2, and jack_fld3 raster datasets sequentially increases the inundation level, and therefore reduces the extent of suitable helicopter landing zones.

7   Close ArcToolbox.

8   Save the ArcMap document as **FL_GISTHS_C6E5.mxd**, with FL being your initials, in **\ESRIPress \GISTHS\MYGISTHS_Work**.

Exercise 6.6
Add the U.S. National Grid

In 2005, the Department of Homeland Security (DHS) recommended that any DHS grant submission reference the use of a nationally defined coordinate system for all spatial referencing, mapping, and reporting. The U.S. National Grid (USNG) is a nonproprietary alphanumeric referencing system derived from the Military Grid Reference System (MGRS). [16] It is a nationally consistent map and spatial grid reference system that seamlessly crosses all jurisdictional boundaries and map scales.

ArcMap enables the overlay of the USNG onto a map layout. Search-and-rescue personnel depend on both GPS-enabled devices and printed map products to find missing persons and return them to safety. The USNG promotes the interoperability between these various types of map products and services by standardizing the plane coordinate system and embedding it in a universal map index.

Convert data frame coordinate system

The USNG is based on the universal transverse Mercator (UTM) coordinate system. The Jackson County geodatabase uses the Mississippi State Plane coordinate system, and needs to be converted to UTM in order to use the USNG.

1   Zoom to the full extent of the JacksonCo_County data layer.

2   Turn on only the FL_jack_hlz1 raster, the Geocode:STATUS2, and the Jackson Co. MEDS layers Hyrdog_WatBodies, Streams, and Streets.

3   Right-click the Layers data frame and select Properties.

4   Click the Coordinate System tab.

5   Expand the Predefined, Projected Coordinate Systems, UTM, NAD 1983 headings and select NAD 1983 UTM Zone 16N.

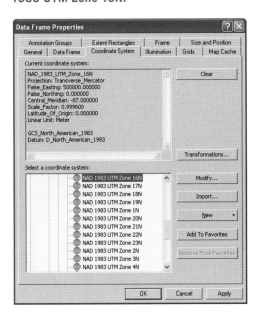

6   Click OK.

The map display is converted to the UTM coordinate system.

Select a grid reference system

1   Right-click the Layers data frame and select Properties.

2   Click the Grids tab.

3   Click New Grid.

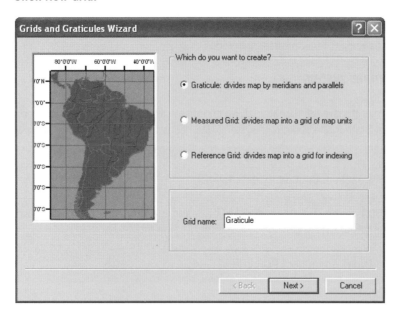

The Grids and Graticules Wizard opens.

4   Click Next three times and Finish once to accept all grid settings.

5   Click Style to open the Reference System Selector and scroll down to select the U.S. National Grid reference system.

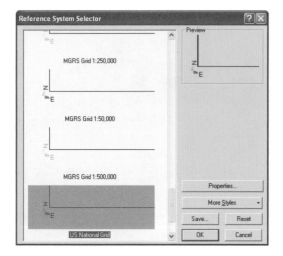

6   Click OK to close the Reference System Selector window.

7   Click OK to close the Data Frame Properties window.

Overlay the USNG onto the map layout

1   Click the layout button ◻ at the lower left corner of the map display window.

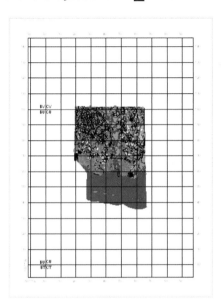

The USNG is overlaid on top of the map of Jackson County.

2   Use the Zoom in tool ⊕ on the Tools toolbar (not on the Layout toolbar!) to zoom to an area where there is a concentration of missing persons.

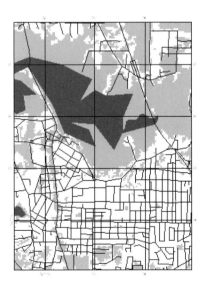

3 Use the Zoom in tool  on the Layout toolbar to zoom to the lower left corner of the map layout.

The coordinate northing and easting grid markers and cell numbers are labeled on the map layout. Any location on a map can be defined using the USNG. Precision and accuracy range from a four-digit coordinate location of a point within a 1,000 meter square, to a 10-digit coordinate location within a 1-meter coordinate square. Each location can be measured with a transparent FGDC Romer Scale grid reader.[17]

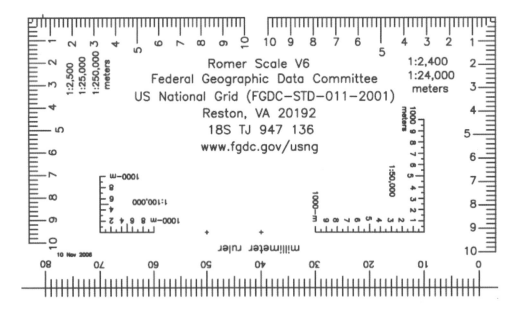

Reading right from the left grid line to the point of location; and then repeating the sequence again by reading up from the lower grid line to the point of location, and dropping the 1-meter level values of 0 results in an eight-digit coordinate location.

In this example, the circled missing person marker is located at 5255 6348. It is:

- 550 meters to the right of grid cell 52
- 480 meters up from grid cell 63

Complete details of the USNG and instructions on how to read and measure this coordinate system are located at www.fgdc.gov/usng.

4 Save the ArcMap document as **FL_GISTHS_C6E6.mxd**, with FL being your initials, in **\ESRIPress \GISTHS\MYGISTHS_Work**.

Summary

You have now completed the chapter on Response. You have learned how the *Respond* Mission Area is activated during a major natural disaster as described in the DHS National Planning Scenario 10. This hypothetical event played out almost exactly as depicted when Hurricane Katrina later struck the Gulf Coast in 2005. The *Respond* Mission Area was activated to meet the DHS search-and-rescue target capability by focusing its efforts on ensuring the safety of missing persons reported to the state emergency management agency. The initial task was to geocode a set of street addresses that were reported as the last known location of missing persons. The next task was to prepare a suitability map to identify available sites for helicopter landing zones in the impacted areas in proximity to the geocoded locations of missing persons. The geoprocessing steps to this process were programmed into a model to automate the sequence of operations involved in the analysis so that it could be executed each time inundation conditions changed along the coast. Upon completion of the analysis, the U.S. National Grid was draped over the map layout to provide a nationally defined coordinate system for spatial referencing, mapping, and reporting, as required by DHS.

Notes

1. DHS National Planning Scenarios, Version 20.1, April 2005, p. vii

2. DHS The National Response Plan-Quick Reference Guide, Version 4.0, May 22, 2006, p. 1

3. Ibid., p. 2

4. Ibid., p. 3

5. Reference forthcoming ESRI Press tutorial book on GIS for First Responders (Mike Price & Anne Johnson)

6. DHS National Planning Scenarios, Version 20.1, April 2005, p. 10-5.

7. DHS National Planning Scenarios, Version 20.1, April 2005, p. 10-2.

8. Ibid., p. 10-5.

9. See Dillon, Patrick, "Berkeley 911", *California Alumni Magazine*, November/December 2005, Vol. 116, No. 4 for a detailed description of how the missing-persons database was devised and implemented. Also see mississippi.deltastate.edu for an overview of how GIS was used extensively in the Katrina response and relief effort.

10. This database is a modified facsimile of the missing-persons reports recorded by the Mississippi Emergency Management Agency during Hurricane Katrina. To ensure the security and privacy of residents reported missing in Jackson County during Hurricane Katrina, all of the names in this sample database are fictitious, and addresses were randomly generated from a database of all possible street addresses in the county.

11. "Identifying Helicopter Landing Zones in Wake of Hurricane Katrina," *ArcNews*, ESRI, Winter 2006/2007, Vol. 28, No. 4, p. 34.

12. IKONOS four-meter resolution multispectral imagery of the area was collected by Space Imaging Inc. and delivered to the MASINT/AGI Node for analysis. Ibid., p. 34. Due to the licensing and release restrictions on this data, for the purpose of this tutorial exercise, fictitious inundation area signatures for three time periods were prepared from publicly available Landsat imagery of Jackson County, Mississippi.

13. slope. (n.d.). Dictionary.com Unabridged (v 1.1). Retrieved July 27, 2007, from Dictionary.com Web site: http://dictionary.reference.com/browse/slope

14. ArcGIS Desktop Help.

15. Ibid.

16. "Introducing the United States National Grid," By Mike Price, Entrada/San Juan Inc., ArcUser Online, http://www.esri.com/news/arcuser/0705/usng1of2.html

17. http://www.fgdc.gov/usng/fgdc-usng-gridreader

Analyzing data for homeland security planning and operations—*Recover*

DHS National Planning Scenario 9: *Natural disaster—major earthquake*

MISSION AREA: *PREVENT*
Event not preventable.

MISSION AREA: *PROTECT*
Retrofit critical infrastructure; conduct response training exercises for emergency operations personnel and population.

MISSION AREA: *RESPOND*
EPA/USCG manage hazardous material spills; American Red Cross delivers emergency medical treatment, shelters and food; Joint Information Center distributes instructions to the public; Urban Search and Rescue deployed.

MISSION AREA: *RECOVER*
DHS Target Capability: Restoration of lifelines
DHS Task: Rec.C.3 Provide energy-related support

7.1 Prepare recover scenario map 1: Restoration of lifelines
USGS ShakeMap Scenario
Add BAUA MEDS Data to new ArcMap Document
Add USGS Faults and ShakeMap data layers
Locate areas of greatest hazard potential
Download SHZP Maps
Intersect high-risk hazard zones
Identify critical infrastructure lifelines in hardest hit areas
Locate local utility outages
Detect knocked-out circuit
Identify full circuit loss of service to parcels
Trace an outage to a portion of a circuit
Isolate partial service disruption to parcels
Identify knock-on effects to critical facilities

Define *recover*, to include implementing immediate actions to restore
essential services after a major disaster or event
Locate areas in highest hazard earthquake zones
Identify lifeline service disruptions to power infrastructure and parcels
Enter damage assessment data and hyperlink photos
Prepare damage assessment status maps and reports

Chapter 7

Analyzing data for homeland security planning and operations: *Recover*

This chapter focuses on using GIS tools and analysis to support the *Recover* Mission Area
as identified in the latest Homeland Security Presidential Directive (HSPD) No. 8 National
Preparedness Goal. DHS defines the critical components of a successful recovery plan as the
"ability to restore essential services, businesses, and commerce; clean up the environment and
render the affected area safe; compensate victims; provide long-term mental health and other
services to victims and the public; and restore a sense of well-being in the community."[1]

While much of the emphasis in HSPD 8 is on prevention, preparedness, and response to major events[2], recovering from a natural or man-made disaster requires a tremendous amount of resources and coordination. HSPD 8 identifies the target capabilities in this mission area as structural damage and mitigation assessment, restoration of lifelines, and economic and community recovery.

Structural damage assessment and mitigation includes conducting damage and safety assessment of civil, commercial, and residential infrastructure, and performing structural inspections and mitigation activities. Providing construction management, technical assistance, and other engineering services are other target capabilities.[3] The intent of this activity is ensuring the safety of buildings and infrastructure before they can be occupied again, as well as providing the necessary information and inspections required for securing relief funds from state and federal agencies.

Restoring lifelines means reestablishing key transportation, communications, and utilities systems to facilitate emergency response activities and the community's return to well-being. This includes the management of clearing and restoration activities (e.g., demolition, repairing, reconstruction), and the removal and disposal of debris.[4]

The target capability of economic and community recovery involves short-term and long-term recovery processes after an incident. These processes include identifying the extent of damage caused by the incident through postevent assessments and determining and providing the support needed for recovery and restoration activities.[5]

In this chapter, you will apply GIS tools and analysis to the restoration of lifelines and damage assessment capabilities of the *Recover* Mission Area in the aftermath of a major earthquake.

Chapter 7: Data dictionary

Layer	Type	Layer Description	Attribute	Description
Berkeley_MEDS.gdb	File Geodatabase	Geodatabase of sample MEDS data layers for Berkeley, California		
power	Feature Dataset			
Distribution_Lines	Feature Class	Distribution lines from substation to transformers		
Parcel_Centroid	Feature Class	Centroid point of parcel		
PowerNet	Feature Class	Geometric network		
PowerNet_Junctions	Feature Class	Junction of geometric network		
Stations	Feature Class	Substations		
Transformers	Feature Class	Transformers		
Transmission_Lines	Feature Class	Transmission lines from power plants to substations		
Berkeley_Parcels	Feature Class	Parcel polygons	INSP_ID	Inspector ID
			INSP_DATE	Inspection date
			NUM_STORIE	Number of stories in structure
			FOOTPRINT_	Sq foot area of footprint of structure
			NUM_RESUNI	Number of residential units in structure
			TYPE_CONST	Type of construction
			PRIM_OCCUP	Primary occupancy
			COND_COLLA	Condition of collapse

Data dictionary continued on next page.

Layer	Type	Layer Description	Attribute	Description
			COND_LEAN	Condition of lean
			COND_RACK	Condition of rack
			COND_GROUN	Condition of ground
			PCNT_DAMAG	Percent damage to structure
			POSTING	Safety posting
			F_ACTION	Future action needed
Berkeley_ StreetCenterLines	Feature Class	Street centerlines		
briov_ls,shp	Shapefile	Landslide Hazard Zone polygons		
oake_lq.shp	Shapefile	Liquefaction Hazard Zone		
oake_ls.shp	Shapefile	Landslide Hazard Zone polygons		
oakw_lq.shp	Shapefile	Liquefaction Hazard Zone		
oakw_ls.sp	Shapefile	Landslide Hazard Zone polygons		
rich_lq.shp	Shapefile	Liquefaction Hazard Zone		
rich_ls.shp	Shapefile	Landslide Hazard Zone polygons		
1734.jpg	Raster Dataset	Postearthquake structural damage image		
1740.jpg	Raster Dataset	Postearthquake structural damage image		
1748.jpg	Raster Dataset	Postearthquake structural damage image		
1768.jpg	Raster Dataset	Postearthquake structural damage image		
1780.jpg	Raster Dataset	Postearthquake structural damage image		
1784.jpg	Raster Dataset	Postearthquake structural damage image		
USGS_BA_Faults	Shapefile	Fault lines	NAME	Name of fault line
USGS_M73_Full_ Hayward_Rodgers_ Faults.shp	Shapefile	Shake Map peak acceleration polygons	VALUE	Level of earthquake shaking acceleration

Exercise 7.1
Prepare recover scenario map 1:

Restoration of lifelines

DHS National Planning Scenario 9:
> Natural disaster—major earthquake

DHS Target Capability:
> Restoration of lifelines

DHS Universal Task:
> Rec.C.3 Provide energy-related support

In this scenario, two catastrophic earthquakes hit a major metropolitan area. DHS National Planning Scenario 9: Natural disaster—major earthquake, is described as follows:

> A 7.5-magnitude earthquake, with a subsequent 8.0-earthquake following, occurs along a fault zone in a major metropolitan area. MM Scope VIII or greater intensity ground shaking extends throughout large sections of the metropolitan area, greatly impacting a six-county region with a population of approximately 10 million people.[6]

The exercises in this chapter return to the San Francisco Bay Area to study the effects of a major earthquake in the region. The Bay Area is within the Pacific Ring of Fire, an area of frequent earthquakes and volcanic eruptions encircling the basin of the Pacific Ocean. Ninety percent of the world's earthquakes and 81 percent of the world's largest earthquakes occur along the Ring of Fire.[7] Extensive efforts and resources are invested throughout the Bay Area to prepare for and reduce the impacts that an event such as this may have on the safety and security of its residents and infrastructure.[8]

The *Recover* Mission Area phase of a homeland security operation includes the assessment of the emergency's impacts as well as the mobilization of recovery personnel to rebuild and restore damaged infrastructure. This National Planning Scenario includes the use of postevent real-time seismic data from the USGS to assess the impact and damage from the earthquakes shortly after they occur. In this exercise, you will use a Bay Area earthquake scenario prepared by the USGS as a planning tool to determine the extent of potential earthquakes forecasted to hit the region in the next 30 years.[9] These scenarios identify which areas are expected to suffer the most extensive damage from shaking and liquefaction, and provide a "best guess" at the level of resulting damage to public infrastructure and private property.

Damage assessments initially focus on the disruption to transportation lines and nodes, power generation and distribution; communications lines; fuel storage and distribution; structures of concern (e.g., dams, levees, nuclear power plants, hazardous materials storage facilities); and structures for provision of essential services (e.g., hospitals and schools typically used as shelters).[10] After initial assessments are complete and the restoration of lifelines are under way, the focus shifts to assessing private property damage at the local community and parcel level to begin moving residents back into their neighborhoods and homes.

USGS ShakeMap scenario

The USGS ShakeMap scenario most closely aligned with the specifications of this National Planning Scenario is the M7.3 Full Hayward & Rodgers Creek Faults. This scenario maps the potential shaking, damage, peak acceleration and velocity, and instrumental intensity of a magnitude 7.3 earthquake along the full extent of the Hayward and Rodgers Creek faults. These faults run northwest/southeast to the east of San Francisco Bay, passing directly through the city of Berkeley.

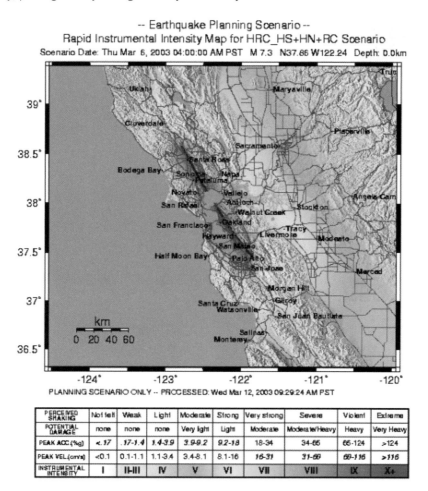

The USGS Earthquake Hazards Program Web site displays graphics of each of the scenario predictions, and provides shapefiles downloadable for use in a GIS.[11]

Add BAUA MEDS data to new ArcMap document

1   Open a new ArcMap Document.

2   Add the BAUA_Boundaries and Hydrography data layers from the **\ESRIPress\GISTHS\MYGISTHS_Work** folder.

3   Turn on only the FL_BAUA_Counties and NHDArea data layers.

4   Zoom to the Bay Area.

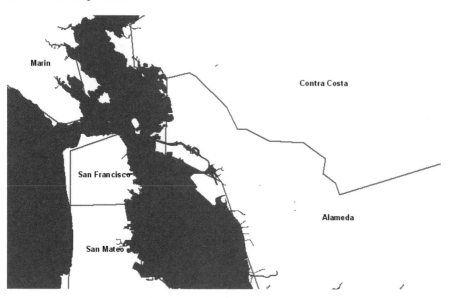

Add USGS Faults and ShakeMap data layers

1   Add the USGS_BA_Faults.lyr and USGS_M73_Full_Hayward_Rodgers_Faults.shp files from the \ESRIPress\GISTHS_C7 folder.

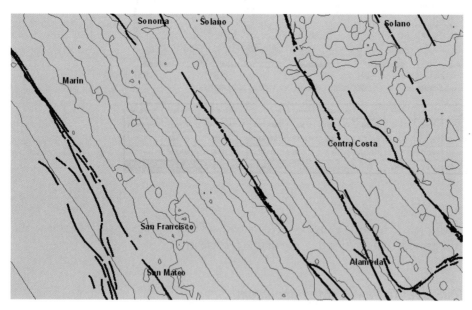

A series of irregular polylines is drawn on the map display, each representing a major fault line in the Bay Area. The polygons resembling contour lines represent an area of peak shaking acceleration of a magnitude 7.3 earthquake along the Hayward and Rodgers Creek faults.

EXERCISES

2 Select By Attributes the faults where "NAME" = 'Hayward fault zone, Northern Hayward section' OR "NAME" = 'Hayward fault zone, Southern Hayward section' OR "NAME" = 'Rodgers Creek fault'.

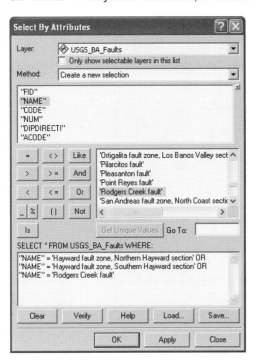

3 Click OK.

4 From the Selection menu, select Zoom to Selected Features.

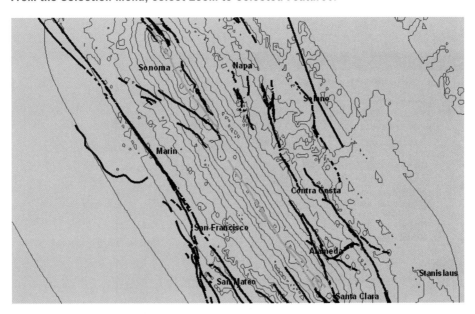

The map display zooms to the selected fault lines to show their geographic extent in the region.

EXERCISES

5 Change the symbology of the USGS_M73_Full_Hayward_Rodgers_Faults data layer to graduated color to show the increasing intensity of the VALUE field indicating the level of shaking acceleration.

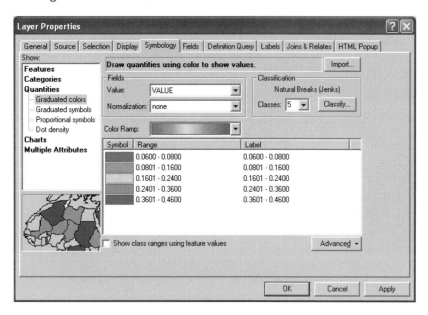

At this point, the lower acceleration values are shown in red and the higher values in green. A conventional ShakeMap reverses these colors to show the highest areas of shake acceleration in red, graduating outward toward yellow and green as the acceleration diminishes.

6 Click the Symbol heading in the legend box.

7 Select Flip Symbols.

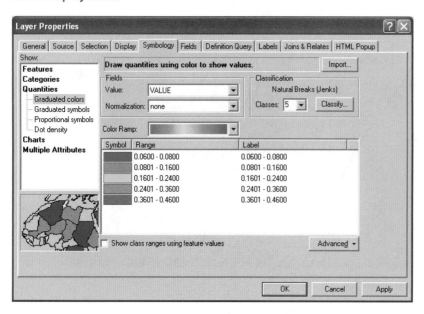

The reversed colors are now ramped to display the more conventional ShakeMap symbology.

8 Click the Symbol heading again and select Properties for All Symbols.

9   Chose No Color for Outline Color, and click OK.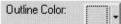

10 Click OK.

11 Drag the ShakeMap data layer below the Hydrography data layer.

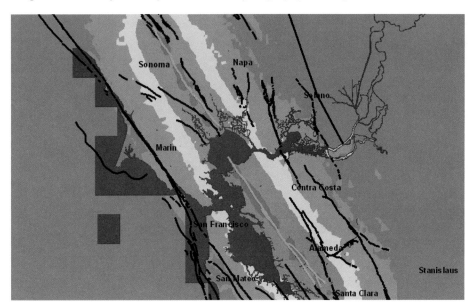

This ShakeMap now shows the extent of the impact of this magnitude earthquake along these two fault lines. The areas in red are hardest hit by this event and, if heavily populated, require immediate attention from first responders to secure the safety and security of people and property. Initial recovery operations to restore power, communications, and transportation lifelines are most critical in these hardest hit areas.

12 Save the ArcMap document as **FL_GISTHS_C7E1.mxd**, with FL being your initials, in **\ESRIPress \GISTHS\MYGISTHS_Work** folder.

Locate areas of greatest hazard potential

The California Geological Survey has compiled a series of shapefiles as part of its Seismic Hazards Zonation Program that show the extent of areas most vulnerable to landslide and liquefaction in the event of a major earthquake. Intersecting these areas with those that experience the greatest peak acceleration pinpoint those locations that are at the greatest risk of catastrophic damage to people and property, including public utilities.

Download SHZP Maps

1   Navigate your Internet browser to http://www.conservation.ca.gov/cgs/shzp/.

2 Click the Interactive Mapping and GIS Data link.

Interactive Mapping and GIS Data

3 Click the Download Data radio button on the left navigation bar.

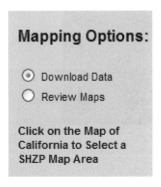

Mapping Options:

⊙ Download Data
○ Review Maps

Click on the Map of California to Select a SHZP Map Area

4 Click the Bay Area SHZP Map Area on the map of California.

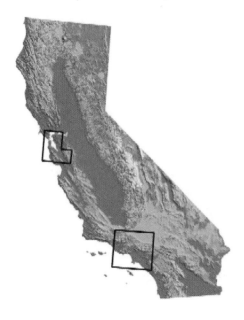

5   Select Berkeley from the Cities drop-down list.

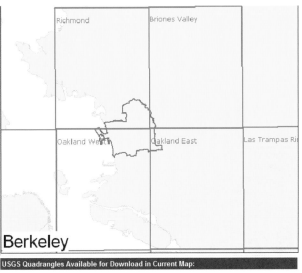

A map of Berkeley is shown on the screen with a USGS Quadrangle map grid overlaid upon it to show the maps available for download. For the sake of this exercise, only the city of Berkeley and the surrounding areas will be downloaded. Compiling the Seismic Hazards Zonation Program maps for the entire BAUA would be a considerable effort, beyond the scope of this exercise, though the processing and analysis steps remain the same.

6   Click Richmond link below the map to download the first file. Register on the Web site to enable access to the download feature.

7   Click   <u>Landslide Zone and Liquefaction data - - - Arcview (.dbf, .shp, .shx) format</u> .

8   Download and extract these files to \ESRIPress\GISTHS\MYGISTHS_Work.

9   Repeat the selection, download, and extraction process for the remaining USGS Quadrangle maps that cover the city of Berkeley:

- **Briones Valley**
- **Oakland West**
- **Oakland East**

10   From ArcCatalog, define the spatial reference for all of the downloaded shapefiles as Geographic Coordinate System North American Datum 1983.

   briov_ls.shp   Note: There are two shapefiles for each quadrangle, _ls for Landslide,
   oake_lq.shp   and lq for Liquefaction; but only one shapefile, _ls, for Briones.
   oake_ls.shp
   oakw_lq.shp
   oakw_ls.shp
   rich_lq.shp
   rich_ls.shp

11 Return to the **\ESRIPress\GISTHS\FL_GISTHS_C7E1.mxd** map document and turn off all data layers except Boundaries group data layer.

12 Add these new Liquefaction and Landslide shapefiles to the map document, and zoom to the extent of these layers.

The map shows the extent of areas mapped for potential landslide and liquefaction in the event of a major earthquake as described in this planning scenario.

13 To facilitate more efficient data processing, use the Merge tool in ArcToolbox to combine all of the _ls files together into one shapefile, **\ESRIPress\GISTHS\MYGISTHS_Work\FL_Landslide_Merge**; and, all of the _lq files together into another, **\ESRIPress\GISTHS\MYGISTHS_Work \FL_Liquefaction_Merge**, with FL being your initials.

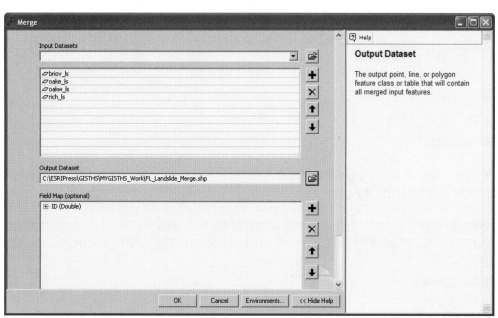

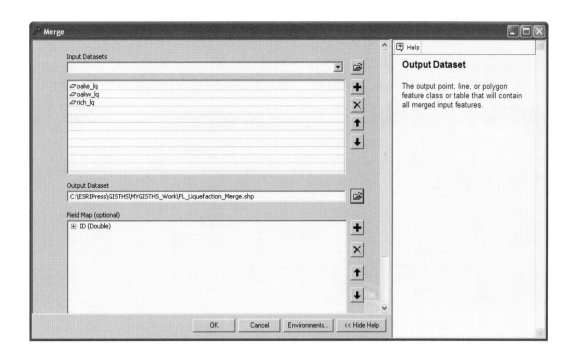

14 Remove all of the individual _ls and _lq shapefiles from the map document.

The map contains a single Landslide layer and single Liquefaction layer covering the extent of the city of Berkeley and surrounding areas. These two critical variables can now be considered when determining the areas of greatest hazard potential from a major earthquake in this area.

15 Export each of these data layers using the same coordinate system as the data frame. Save the output feature classes as **FL_Landslide_UTM** and **FL_Liquefaction_UTM**, with FL being your initials, in **\ESRIPress\GISTHS\MYGISTHS_Work**.

Using the Union tool combines these two different shapefiles into one data layer covering the full extent of area impacted by these hazards.

16 From ArcToolbox, expand Analysis Tools, Overlay and select Union.

17 Select FL_Landslide_UTM and FL_Liquefaction_UTM as the Input Features; and save the output feature class as **FL_Landslide_Liquifaction_Union** in **\ESRIPress\GISTHS\MYGISTHS_Work**.

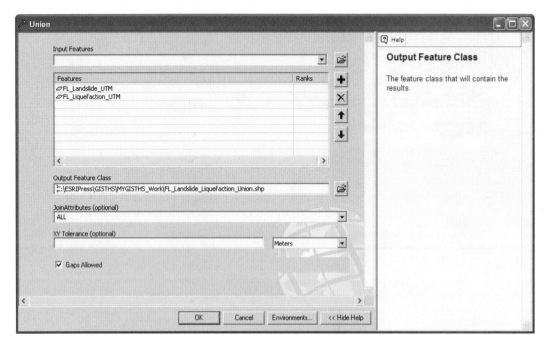

18 Click OK.

19 Close the Union window when complete.

The new feature is added to the map showing the combined extent of both data layers.

20 Remove FL_Landslide_Merge, FL_Landslide_UTM, FL_Liquefaction_Merge, and FL_Liquefaction_
UTM from the map document.

21 Save the ArcMap document.

Intersect high-risk hazard zones

1 Turn on the USGS_BA_Faults, USGS_M73_Full_Hayward_Rodgers_Faults, and Hydrography data
layers.

2 Clear selected features.

3 Select By Attribute from USGS_M73_Full_Hayward_Rodgers_Faults where the peak acceleration
value is greater than 0.36.

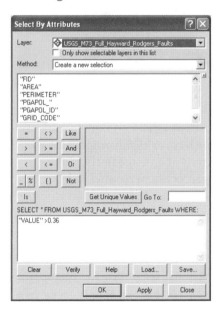

4 Click OK and zoom to the selected features.

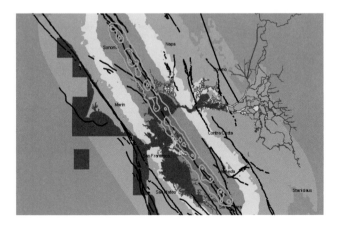

Only the areas with the greatest peak acceleration are selected and highlighted.

5   Zoom to the extent of FL_Landslide_Liquifaction_Union.

The map display shows the locations in the East Bay that experience the greatest peak acceleration during a major earthquake event, and are most vulnerable to liquefaction and landslide. These areas in the highest peak acceleration zones are critically important to recovery personnel as they initially assess the level and extent of damage in the wake of a major earthquake. Refining the analysis further enables recovery personnel to hone in more closely to those areas where the combined impacts of peak acceleration, landslide, and liquefaction exist in common.

6   From ArcToolbox, expand Analysis Tools, Overlay and select Intersect.

7   Select FL_Landslide_Liquifaction_Union and USGS_M73_Full_Hayward_Rodgers_Faults as the Input Features; and save the output feature class as **FL_Land_Liq_PAccel**, with FL being your initials, in **\ESRIPress\GISTHS\MYGISTHS_Work**.

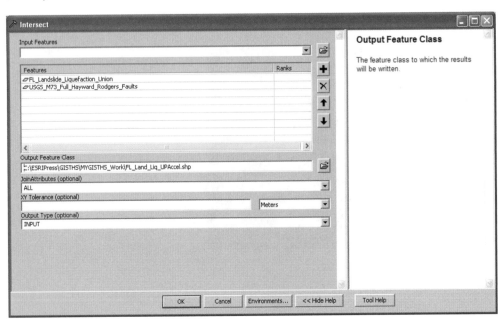

8   Click OK.

9   Close the Intersect window when complete.

10   Clear selected features and turn off FL_Landslide_Liquifaction_Union to view only the new
     intersect data layer.

Only those areas of landslide and liquefaction risk in the highest peak acceleration zone
are included. These are the areas where initial recovery efforts are to be focused, as they are
the locations that are most vulnerable to the greatest damage from a major earthquake of
this magnitude.

11   Save the ArcMap document.

Identify critical infrastructure lifelines in hardest hit areas

Restoring lifelines is the reestablishment of key transportation, communications, and utilities systems to
facilitate emergency-response activities and the community's return to well-being. The success of this
recovery effort depends heavily on how the operations of these infrastructure systems are managed before a
major catastrophic event occurs. If they are already integrated into a functional enterprise GIS system, such
as ArcFM[12], then inventory and recovery of the system components is more effectively accomplished. Most
utility companies are well on their way to integrating their systems operations and maintenance activities into
an enterprise GIS environment. The development of this type of facilities management system is beyond the
scope of this tutorial. However, this exercise is designed to support recovery efforts by emergency operations
personnel working within an enterprise GIS system for facilities management, focusing on the recovery of gas
and electrical service after a major catastrophic event that impacts utility services in the most vulnerable
locations.[13]

Damage to high-pressure gas transmission lines poses a tremendous risk to communities due to the volatile
nature of this infrastructure. The combined impacts of landslide, liquefaction, and earthquake acceleration
can result in secondary damage to these lines, to include rupture and subsequent explosion and fire that

threaten surrounding communities and infrastructure. Sections of gas lines that are located in these high-risk zones can be identified and tagged within a GIS for immediate response and recovery action in the event of an earthquake of significant magnitude.

1   Add **BAUA_Gas_Transmission_Pipelines.lyr** from **\ESRIPress\GISTHS\GISTHS_C7** to the map document and zoom to its full extent.

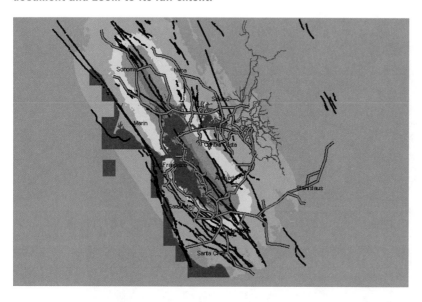

This dataset shows the extent of the high-pressure gas transmission lines in and around the Bay Area.

2   Zoom to the extent of **FL_Land_Liq_PAccel.**

There are segments of this pipeline that run through those areas vulnerable to landslide, liquefaction, and peak ground acceleration. These segments are in danger of rupture, and subsequent explosion and fire, and require immediate attention in the event of an earthquake of this magnitude.

3   Intersect BAUA_Gas_Transmission_Pipelines and FL_Land_Liq_PAccel to single out those segments of the pipeline that are at the greatest risk of rupture. Save this new output feature call as **FL_BAUA_HighRisk_GasPipelines.shp**, with FL being your initials, in **\ESRIPress\GISTHS \MYGISTHS_Work**.

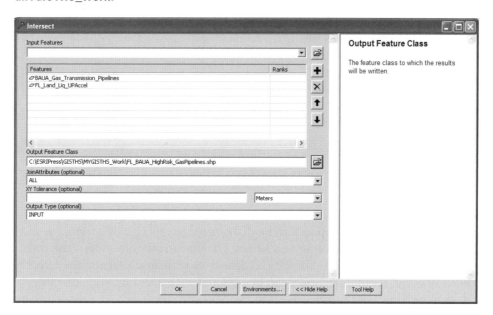

4   Close the Intersect window when complete.

5   Change the color and thickness of this polyline feature to distinguish it clearly in the map display, and label the pipeline segments using the GP_ID field.

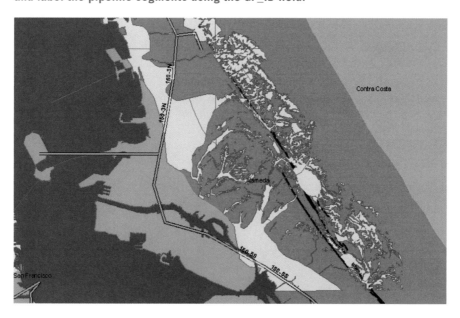

EXERCISES

6 Open the Attributes of FL_BAUA_HighRisk_GasPipelines table to view the segments of the gas
pipeline that are at the greatest risk of rupture in the event of an earthquake of this magnitude.

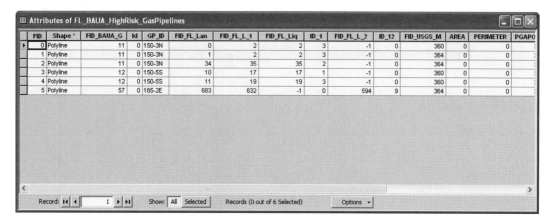

FID	Shape	FID_BAUA_G	Id	GP_ID	FID_FL_Lan	FID_FL_L_1	FID_FL_Liq	ID_1	FID_FL_L_2	ID_12	FID_USGS_M	AREA	PERIMETER	PGAP0
0	Polyline	11	0	150-3N	0	2	2	3	-1	0	360	0	0	
1	Polyline	11	0	150-3N	1	2	2	3	-1	0	364	0	0	
2	Polyline	11	0	150-3N	34	35	35	2	-1	0	364	0	0	
3	Polyline	12	0	150-5S	10	17	17	1	-1	0	360	0	0	
4	Polyline	12	0	150-5S	11	19	19	3	-1	0	360	0	0	
5	Polyline	57	0	185-2E	683	632	-1	0	594	9	364	0	0	

Record: 1 ▸▸| Show: All Selected Records (0 out of 6 Selected) Options ▾

This map and table in the form of a report can be distributed to emergency services personnel
as they embark on the recovery mission to secure and stabilize utility lifelines in the wake
of a major catastrophic earthquake. As the USGS generates real-time shake maps directly
after an earthquake event, this information can be updated with accurate and current ground
acceleration data to redefine the high risk areas as needed, and hence, reselect those segments
of the pipeline that are at the greatest risk of rupture.

7 Close the attribute table, and turn off the FL_BAUA_HighRisk_GasPipelines, BAUA_Gas_
Transmission_Pipelines, and FL_Land_Liq_PAccel data layers.

8 Save the ArcMap document.

Locate local utility outages

Local electrical grid systems suffer from isolated service outages and disruptions as a regular course of
operations. After a major catastrophic earthquake, the outages may be extensive along fault lines due to
damaged poles, transformers, and wires caused by falling debris and fire. The critical steps in an outage
management system include determining:

- the location of the actual failure point (or fault) or points;
- the location of the disconnecting device or devices;
- the customers associated with each of the events.[14]

1　Add the Berkeley_Parcels[15] feature from the Berkeley_MEDS geodatabase located in the **\ESRIPress\GISTHS_C7 folder**, and zoom to the extent of this layer.

The map shows that almost the entire city of Berkeley sits within the area of greatest peak acceleration, and one of the fault lines runs directly through the eastern portion of the city.

2　Add the "power" feature dataset from the Berkeley_MEDS geodatabase located in the **\ESRIPress \GISTHS_C7** folder, and zoom to the extent of these layers.

The "power" feature dataset contains a set of features that make up the electrical grid system for a portion of the city of Berkeley. These features include the following data layers:

- Transmission lines: Transmit power from generating plants located outside of the city.
- Substations: Receive the power generation feed from the power plants via the transmission lines, and step down the power to lower levels.
- Distribution lines: Disperse power to the transformers along circuits originating from the substations.
- Transformers: Rest on poles and feed power to clusters of individual parcels.
- Parcel centroid: The center point of the parcel within which it resides.
- PowerNet junctions: The junction points that provide connectivity through the utility network.

3   Remove Parcel Centroid from the Table of Contents and turn off the USGS_M73_Full_Hayward_ Rodgers_Faults data layer.

4   Zoom to the area of the city where the fault line transects.

5   Change the symbology of the "power" features to show up more clearly on the map.

6   Label the Distribution Lines using the cir_num field.

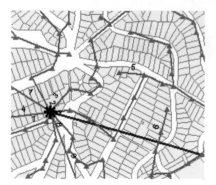

Detect knocked-out circuit

If circuit no. 5 was knocked out by a jolt at the substation, ArcMap is able to identify the parcels that would be affected by this loss of service.

1   **Select by attributes Distribution Lines where cir_num equals 5.**

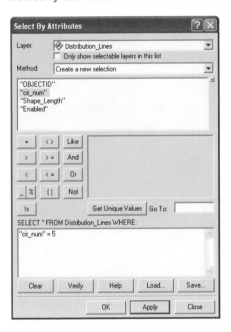

2   **Click OK.**

The entire circuit is selected to show the extent of the area this portion of the grid services. Along the distribution lines lay transformers resting on poles that feed power to clusters of parcels. ArcMap can pinpoint the transformer poles, and the parcels that are serviced along this circuit to identify those customers experiencing power outages as a result of this event.

3   Select by location the transformers that intersect with the selected features of the distribution lines.

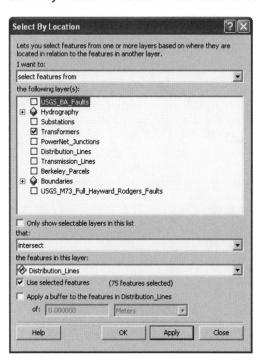

4   Click OK.

The transformers along this knocked out circuit are now selected.

5   Export the selected transformers to a new shapefile, **FL_Circuit5_TP.shp**, with FL being your initials, in **\ESRI\GISTHS\MYGISTHS_Work**.

6   Clear selected features.

Identify full circuit loss of service to parcels

The parcels attribute table contains a field that relates each parcel to a transformer pole (TP). Each transformer pole is related to a circuit, so when a transformer is rendered inoperative, the related parcels along that circuit can be identified.

1 Join **FL_Circuit5_TP** to **Berkeley_Parcels** using **SUM_TP_NUM** and **TP_num** as the join fields.

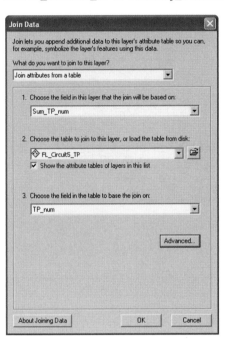

2 Click **OK**, and **Yes** to create an index, if asked.

3 Select by attributes the Berkeley parcels where **FL_Circuit5_TP.cir_num** equals 5.

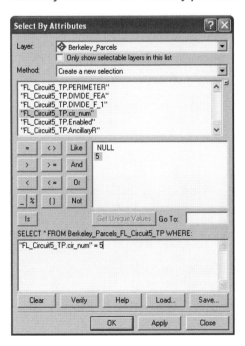

4   Click OK.

All of the parcels that experience a service outage as a result of circuit no. 5 getting knocked out by a jolt at the substation are now selected.

5   Generate a report listing the parcel numbers and street addresses of only the selected sites experiencing a power outage as a result of this event.

Berkeley Power Outage Parcels: Circuit #5

Parcel #	Address	Street
061 25889007.	2338	SPRUCE ST
061 256340010	2274	MARIPOSA AVE
061 255610054	2335	SPRUCE ST
061 256470056	2337	SUTTER ST
061 255960034	2271	MARIPOSA AVE
061 259000090	2161	SANTA BARBARA RD
063 297220053	2224	EUCLID AVE
061 258630089	2190	SHATTUCK AVE
061 258770074	2238	OXFORD ST
063 297440073	2253	KEITH AVE
063 29743033.	2257	KEITH AVE
061 255550014	2268	KEITH AVE
061 255580018	2319	SPRUCE ST
061 258830088	2281	OXFORD ST
061 256300065	2258	MARIPOSA AVE
063 29871007.	2297	EUCLID AVE
063 297230084	2268	EUCLID AVE
061 255990023	2255	MARIPOSA AVE
061 256380375	3246	DEL NORTE ST
061 255840019	3354	LOS ANGELES AVE
061 256450021	2284	MARIPOSA AVE
061 255760097	3444	LOS ANGELES AVE
061 258780093	3353	LOS ANGELES AVE
061 258750013	2270	OXFORD ST
061 258840068	2302	SPRUCE ST
061 255510456	2318	EUCLID AVE
061 25861009.	2180	SHATTUCK AVE
061 258770022	2187	SHATTUCK AVE
063 297450166	2246	CRAGMONT AVE
061 256360889	2242	MARIPOSA AVE
063 297400048	2273	KEITH AVE
061 255530026	2288	KEITH AVE

This map and report can be transmitted to emergency response and service operations personnel identifying the locations of the parcels experiencing power outages as a result of this event. Resources can then be accurately estimated and harnessed to restore service to these parcels.

Trace an outage to a portion of a circuit

In this scenario, multiple transformer poles scattered throughout the area have been severely damaged as a result of a jolt occurring during the earthquake. Damage may occur at one or more locations along a circuit that may not affect the service of the entire circuit. The ArcMap extension, Utility Network Analyst, enables tracing along a geometric network to identify paths the length of the entire the system, or portions of it, that are connected and experience outages as a result of partial disruptions along the line.

1   Clear all selected features and turn off FL_Circuit5_TP.

2   From the View menu, select Toolbars and Utility Network Analyst.

3   Select PowerNet as the Network to use in this analysis.

4   Click the Analysis drop-down menu and select Options.

5   Click the Results tab and click Selection and uncheck Edges so only Junctions is checked.

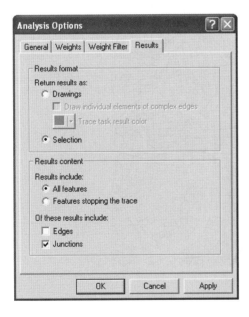

6   Click OK.

7   Label Transformers using TP_num as the label field.

8 Select by attributes the transformers where transformer numbers (TP_Num) equal 155, 211, 259, 305 or 374. These are the transformer poles along the fault line that have been reported damaged in the aftermath of the earthquake.

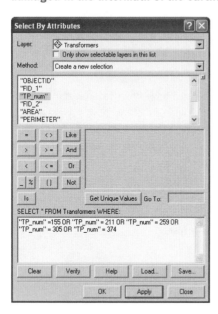

9 Click OK and zoom to the selected features.

These damaged poles are now selected in the map display.

10 Click the Add Junction Flag Tool ⌐ from the Utility Network Analyst toolbar.

11   Click each of these poles to place a green square junction flag over each marker.

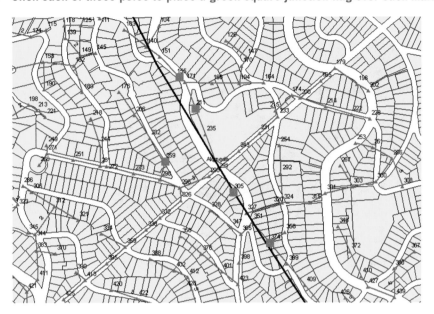

12   Click the Flow drop-down menu and select Display Arrows For Only Distribution Lines.

13   Select Display Arrows from the same menu.

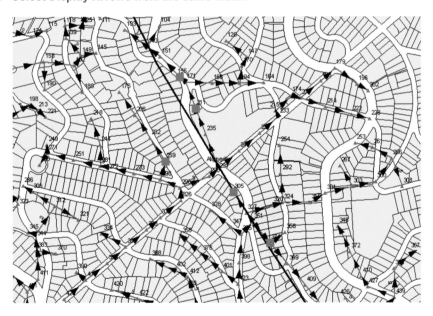

The arrows show the direction of flow of the circuit network. Tracing the "downstream" path from the damaged transformer poles along the network will identify other transformers in the circuit that are rendered inoperable by the break in the circuit flow. The parcels related to these transformers can then be located to isolate only those residents experiencing power outages resulting from this partial disruption of service.

14   Select Trace Downstream as the Trace Task on the Utility Network Analyst toolbar.

15 Click the Solve button ⚹ , and zoom to the selected features.

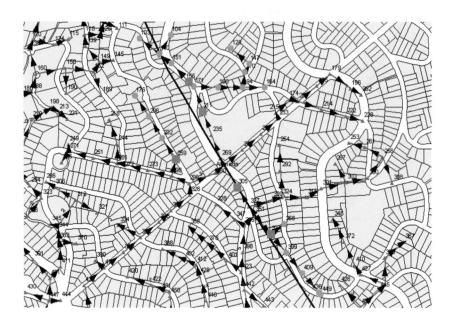

All of the transformer poles located beyond the damaged poles are selected and highlighted. The parcels serviced by the damaged and inoperable transformers can now be identified to direct recovery efforts to restore power to these residents.

16 Export the selected transformers to a new shapefile, **FL_DS_TP.shp**, with FL being your initials, in **\ESRI\GISTHS\MYGISTHS_Work**.

17 Clear selected features.

18 Click the Analysis drop-down menu and select Clear Flags to remove the flags from the map display.

19 Click the Flow drop-down menu and select Display Arrows to toggle off the flow arrows.

Isolate partial service disruption to parcels

1 Right-click Berkeley_Parcels and select Joins and Relates, Remove Joins, FL_Circuit5_TP.

2   Join FL_DS_TP to Berkeley_Parcels to relate the damaged and inoperative transformer poles to the parcels they service.

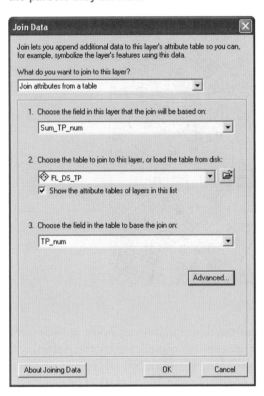

3   Click OK, and Yes to create an index, if asked.

4   Select by attributes the Berkeley parcels where FL_DS_TP.cir_num IS NOT NULL.

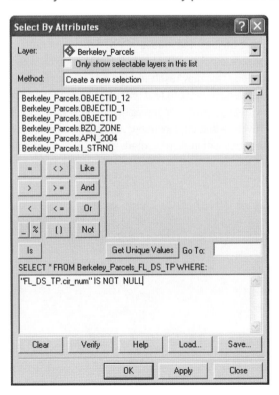

5   Click OK.

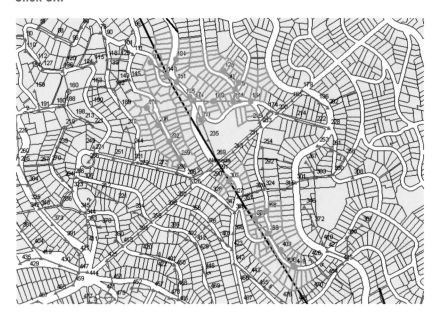

Only those parcels serviced by the damaged and inoperable transformer poles are selected. This process isolates only those within the service area that are impacted by the partial disruption of service resulting from damage to poles along the fault line. Recovery personnel can more effectively concentrate restoration efforts on these selected residents to return them to full power in the aftermath of a major disaster.

6   Save the map document when complete.

Identify knock-on effects to critical facilities

Electrical service disruptions have knock-on effects to other utilities and infrastructure dependent on these power systems. Water pumping stations that experience a power outage will suffer loss of, or decline in, operations once their backup power supplies are depleted. This may result in a drop in water pressure throughout the water distribution system, affecting the delivery of water to firefighters and emergency shelters. Knowing the backup capacity of these facilities greatly assists recovery personnel as they allocate resources to restore these services to normal operation levels. Additional locations, such as schools, hospitals, and churches that could serve as shelter and recovery facilities within disrupted areas may also be affected by the loss of power along a damaged circuit.

1   Select Berkeley_Parcels by attributes from the current selection where the USEDSC equals CHURCHES, HOSPITALS, OWNED BY PUBLIC UTILITY or SCHOOLS.

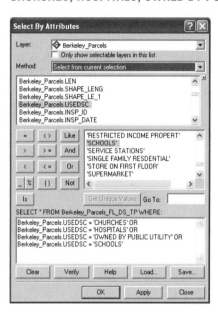

2   Click OK.

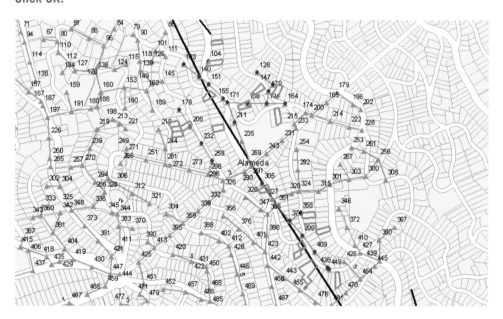

Only those parcels in the disrupted area that are in these use categories are selected.

3   Export the selected parcels to a new shapefile, **FL_DS_CF.shp**, with FL being your initials, in **\ESRI\GISTHS\MYGISTHS_Work**.

4   Clear selected features, and change the symbology of this new shapefile to show the different categories of use of these parcels.

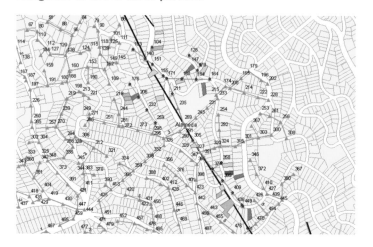

CHURCHES
HOSPITALS
OWNED BY PUBLIC UTILITY
SCHOOLS

Real world parcel databases usually contain specific landowner name and contact information to enable precise and exact identification of impacted facilities. Additionally, public utilities often have integrated spatial and operational network databases that link distinct yet dependent services—such as electric, gas, and water—in a comprehensive schema that can identify through relational associations which services may be affected by a disruption within a specific utility network. Also keep in mind that similar analysis of utility network disruptions can be done with gas, water, and communications network databases as was performed here on the electrical grid network.

5   Save the map document when complete.

YOUR TURN

Create a new map layout and report showing the parcels containing critical facilities that are impacted by a partial disruption in electrical service resulting from a major earthquake.

Analyzing data for homeland security planning and operations—*Recover*

DHS National Planning Scenario 9: *Natural disaster— major earthquake*

MISSION AREA: *PREVENT*
Event not preventable.

MISSION AREA: *PROTECT*
Retrofit critical infrastructure; conduct response training exercises for emergency operations personnel and population.

MISSION AREA: *RESPOND*
EPA/USCG manage hazardous material spills; American Red Cross delivers emergency medical treatment, shelters and food; Joint Information Center distributes instructions to the public; Urban Search and Rescue deployed.

MISSION AREA: *RECOVER*
DHS Target Capability: Structural damage assessment and mitigation.
DHS Task: Rec.C.2.3.1 Assessment of damage to structures, public works, and infrastructure.

7.2 **Prepare recover scenario map 2:**

Damage assessment
Identify parcels in greatest hazard zones
Map damage assessment designation
Enter damage assessment data
Review parcel recovery status maps
Hyperlink photos to maps

Exercise 7.2
Prepare recover scenario map 2:

Damage assessment

DHS National Planning Scenario 9:
> Natural disaster—major earthquake

DHS Target Capability:
> Structural damage assessment and mitigation

DHS Universal Task:
> Rec. C.2. 3.1 Postincident assessment of structures, public works, and infrastructure

Identify parcels in greatest hazard zones

After the restoration of lifelines is under way, the recovery effort shifts to assessing damage to individual structures to enable residents to move back into their homes. Recovery personnel shall direct their initial efforts to those parcels located in the zones with the highest hazard potential.

1   Clear selected features, and turn off all data layers except FL_Land_Liq_PAccel, Berkeley_ Parcels, NHDArea, FL_BAUA_Counties, and USGS_M73_Full_Hayward_Rodgers_Faults.

2   Add the Berkeley_StreetCenterLines feature class from the Berkeley_MEDS geodatabase in the \ESRIPress\GISTHS_C7 folder.

3   Change the symbology of the parcel data layer to Gray 10%; and the streets to Black width 1.

4   Select By Attribute from USGS_M73_Full_Hayward_Rodgers_Faults where the peak acceleration value is greater than or equal to 0.46. Be sure to change the Method to Create a new selection.

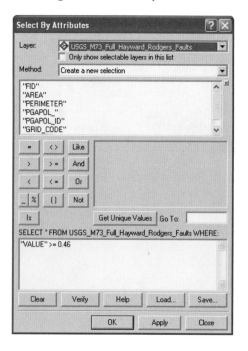

5   Click **OK** and zoom to the extent of the selected features.

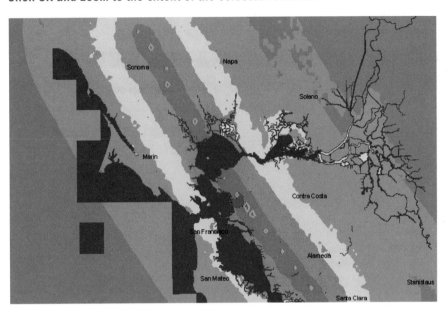

6   Zoom to the diamond shape polygon in **North Berkeley**.

7   Change the symbology of FL_Land_Liq_PAccel to Cherrywood Brown, and increase its transparency to 50%.

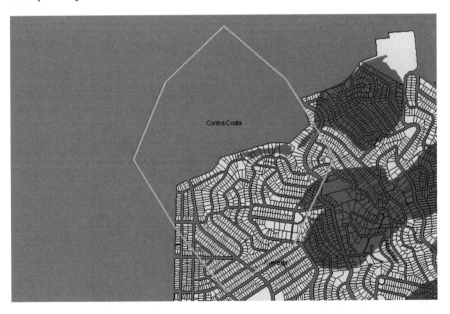

The map display shows those parcels in this greatest peak acceleration zone that are most vulnerable to severe damage from liquefaction and/or landslide in the event of a major earthquake. Recovery personnel are directed to these high-risk parcels for initial damage assessment to determine habitability of the structures and stability of the surrounding landscape.

8 From ArcToolBox, clip the features from FL_Land_Liq_PAccel by the selected features in USGS_ M73_Full_Hayward_Rodgers_Faults. Output the feature class as **FL_Land_Liq_PAccel_GE46.shp**, with FL being your initials, in **\ESRI\GISTHS\MYGISTHS_Work**.

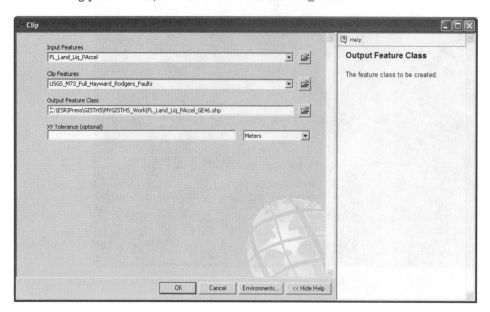

9 Click OK and close the Clip window when complete.

A data layer is added to the map containing only the areas of landslide and/or liquefaction hazard that are within the zone of highest peak acceleration.

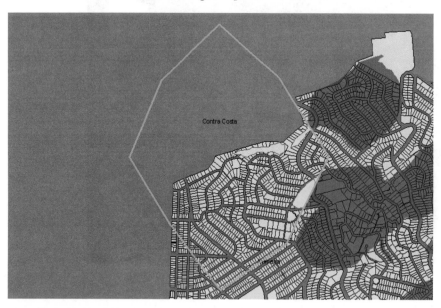

10 Select By Location the features from Berkeley_Parcels that intersect with the FL_Land_Liq_ PAccel_GE46 data layer.

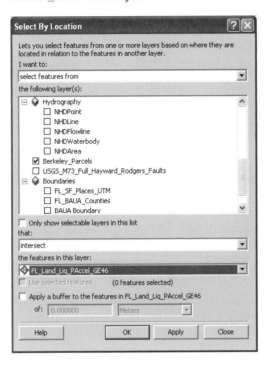

11 Click OK.

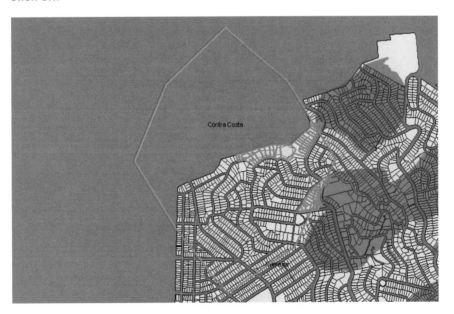

All of the high-risk parcels where this potential hazard of landslide and/or liquefaction exists are selected and highlighted.

12 Open the Berkeley_Parcels attribute table and view only the selected records.

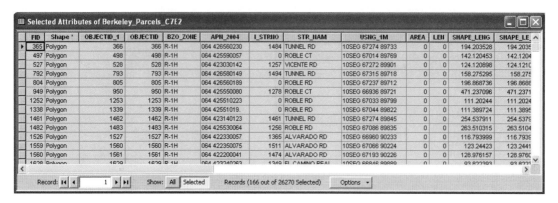

There are 166 parcels where this potential hazard of landslide and/or liquefaction exists.

13 Click Options and select Report to create a report listing the APN_2004, I_STRNO, STR_NAM, and USNG_1M fields of these parcels.

14 Add Berkeley parcels: Rapid Damage Assessment as the title of the report.

15 Add a date to the report as a subtitle.

Berkeley Parcels: Rapid Damage Assessment

Date: 08/16/07

Parcel No.	Address	Street	USNG_1M
048H760700044	1500	ALVARADO RD	10 SEG 67125 90292
048H760720049	1524	ALVARADO RD	10 SEG 67041 90283
048H760780020	1490	ALVARADO RD	10 SEG 67179 90291
048H766370038	1851	ALVARADO RD	10 SEG 66961 90279
048H766380037	1345	ALVARADO RD	10 SEG 66899 90273
062290310144	1688	VINCENTE AVE	10 SEG 63133 94994
062290310033.	1744	THE ALAMEDA	10 SEG 63268 95010
062290320017	1754	THE ALAMEDA	10 SEG 63240 95005
062290320096	0	VINCENTE AVE	10 SEG 63110 94988
062290330075	1690	VINCENTE AVE	10 SEG 63100 94975
062290330079	1758	THE ALAMEDA	10 SEG 63216 94985
062290340002.	0	THE ALAMEDA	10 SEG 63194 95012
062290340081	1696	VINCENTE AVE	10 SEG 63085 94971
062290350045	1764	THE ALAMEDA	10 SEG 63192 94969
062290360015	0	THE ALAMEDA	10 SEG 63165 95000
062290370076	1768	THE ALAMEDA	10 SEG 63186 94948
062290390046	1685	VINCENTE AVE	10 SEG 63154 94963
062291650029	1871	ARLINGTON AVE	10 SEG 63628 94823
062291660031	0	ARLINGTON AVE	10 SEG 63626 94842
062292100046	3197	THOUSAND OAKS BLVD	10 SEG 63563 94680
062292100077	1869	SAN FERNANDO AVE	10 SEG 63536 94752
062292110063	1874	ARLINGTON AVE	10 SEG 63581 94802
062292110088	1861	SAN FERNANDO AVE	10 SEG 63547 94765
062292110170	3185	THOUSAND OAKS BLVD	10 SEG 63542 94679
062292120028	1857	SAN FERNANDO AVE	10 SEG 63553 94760
062292120040	1876	ARLINGTON AVE	10 SEG 63578 94783
062292120174	3171	THOUSAND OAKS BLVD	10 SEG 63515 94671
062292130010	1888	ARLINGTON AVE	10 SEG 63573 94770
062292130068	3157	THOUSAND OAKS BLVD	10 SEG 63483 94669
062292130089	1849	SAN FERNANDO AVE	10 SEG 63564 94792
062292140054	3149	THOUSAND OAKS BLVD	10 SEG 63465 94661
062292140067	1890	ARLINGTON AVE	10 SEG 63568 94757
062292150026	1894	ARLINGTON AVE	10 SEG 63564 94741
062292150057	1891	SAN FERNANDO AVE	10 SEG 63491 94687
062292160014	1887	SAN FERNANDO AVE	10 SEG 63501 94696
062292160008.	1898	ARLINGTON AVE	10 SEG 63563 94726
062292170032	1902	ARLINGTON AVE	10 SEG 63557 94712
062292170084	1881	SAN FERNANDO AVE	10 SEG 63515 94705
062292180019	1877	SAN FERNANDO AVE	10 SEG 63525 94720
062292180099	1906	ARLINGTON AVE	10 SEG 63576 94698
062292190021	1873	SAN FERNANDO AVE	10 SEG 63529 94737
062292300076	1740	ARLINGTON AVE	10 SEG 63541 95128
062292300139	1796	ARLINGTON AVE	10 SEG 63449 95020
062292300492	1734	THE ALAMEDA	10 SEG 63310 95017
062292300544	1740	THE ALAMEDA	10 SEG 63292 95011

This map and report are critical documents that recovery personnel can use to initially identify those parcels that may have sustained the greatest damage in the event of a major earthquake and require rapid onsite assessment to determine safety and habitability.

16 Save the report as **FL_GISTHS_C7E2_R1.rdf**, with FL being your initials, in **\ESRIPress\GISTHS\ MYGISTHS_Work**.

17 Close the table, and save the FL_GISTHS_C7E2.mxd document.

Map damage assessment designation

In the event of a major earthquake, response and recovery personnel are on the scene as soon as possible to ensure the safety and security of people and property. Assessment personnel may include professional and/or volunteer engineers and architects inspecting properties and structures for damage. Onsite field inspections have been greatly enhanced by the use of handheld electronic data entry devices equipped with technology such as ESRI's ArcPad mobile GIS application where integrated GPS receivers and digital assessment forms are linked to GIS parcel maps and aerial imagery.

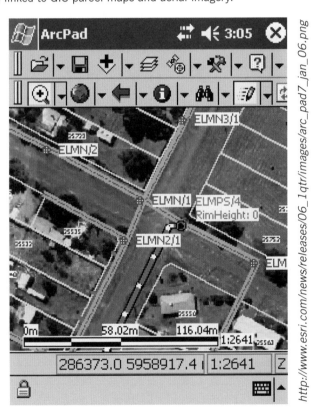

http://www.esri.com/news/releases/06_1qtr/images/arc_pad7_jan_06.png

Though the development of such systems is beyond the scope of this tutorial, this exercise simulates the use of a damage assessment geodatabase that can be used as a mobile onsite data entry tool. Future development of the ALLHAZ application tool discussed in chapter 5 will also greatly enhance the coordinated efforts of onsite recovery personnel as they identify and assess damage to public and private property within a common operating enterprise environment.

The Applied Technology Council (ATC) has developed a series of data entry forms widely accepted and used for onsite rapid and detailed damage assessment.[16] The Rapid Evaluation Safety Assessment Form includes fields that describe the building and indicate the extent of damage sustained from the event. Upon assessment, the structure is then posted using one of a series of ATC placards indicating that it has been deemed either safe for lawful occupancy (green), placed under restricted use (yellow), or marked unsafe for occupancy (red). Further actions are then noted, indicating if the parcel requires barricading, and/or a more detailed structural or geotechnical evaluation.

ATC-20 Rapid Evaluation Safety Assessment Form

Inspection
Inspector ID: _____
Affiliation: _____
Inspection date and time: _____ ☐ AM ☐ PM
Areas inspected: ☐ Exterior only ☐ Exterior and interior

Building Description
Building name: _____
Address: _____
_____
Building contact/phone: _____
Number of stories above ground: ___ below ground: ___
Approx. "Footprint area" (square feet): _____
Number of residential units: _____
Number of residential units not habitable: _____

Type of Construction
☐ Wood frame ☐ Concrete shear wall
☐ Steel frame ☐ Unreinforced masonry
☐ Tilt-up concrete ☐ Reinforced masonry
☐ Concrete frame ☐ Other: _____

Primary Occupancy
☐ Dwelling ☐ Commercial ☐ Government
☐ Other residential ☐ Offices ☐ Historic
☐ Public assembly ☐ Industrial ☐ School
☐ Emergency services ☐ Other: _____

Evaluation
Investigate the building for the conditions below and check the appropriate column.

Observed Conditions: — Minor/None — Moderate — Severe
Collapse, partial collapse, or building off foundation ☐ ☐ ☐
Building or story leaning ☐ ☐ ☐
Racking damage to walls, other structural damage ☐ ☐ ☐
Chimney, parapet, or other falling hazard ☐ ☐ ☐
Ground slope movement or cracking ☐ ☐ ☐
Other (specify) _____ ☐ ☐ ☐

Estimated Building Damage (excluding contents)
☐ None ☐ 0–1% ☐ 1–10% ☐ 10–30% ☐ 30–60% ☐ 60–100% ☐ 100%

Comments: _____

Posting
Choose a posting based on the evaluation and team judgment. *Severe* conditions endangering the overall building are grounds for an Unsafe posting. Localized *Severe* and overall *Moderate* conditions may allow a Restricted Use posting. Post INSPECTED placard at main entrance. Post RESTRICTED USE and UNSAFE placards at all entrances.

☐ **INSPECTED** (Green placard) ☐ **RESTRICTED USE** (Yellow placard) ☐ **UNSAFE** (Red placard)

Record any use and entry restrictions exactly as written on placard: _____

Further Actions Check the boxes below only if further actions are needed.
☐ Barricades needed in the following areas: _____
☐ Detailed Evaluation recommended: ☐ Structural ☐ Geotechnical ☐ Other: _____
☐ Other recommendations: _____
Comments: _____

The Berkeley parcel geodatabase has been appended for use in this tutorial with fields for each of these indicators. Below is a list of the data entry codes used to indicate the status of the inspected parcels.

Type of Construction:

WF	Wood Frame
SF	Steel Frame
TC	Tilt-up concrete
CF	Concrete frame
CS	Concrete shear wall
UM	Unreinforced masonry
RM	Reinforced masonry
OT	Other

Primary Occupancy:

D	Dwelling
OR	Other residential
PA	Public Assembly
ES	Emergency services
C	Commercial
OF	Offices
I	Industrial
G	Government
H	Historic
S	School
OT	Other

Evaluation:

MN	Minor
MD	Moderate
SV	Severe

Posting:

G	Green: Inspected
Y	Yellow: Restricted
R	Red: Unsafe

Further Actions:

N	None
B	Barricades needed
DS	Detailed structural evaluation needed
DG	Detailed geotechnical evaluation needed
DO	Detailed other evaluation needed

For this exercise, a set of parcels in the high-risk zone has been preinspected using these assessment codes. There are parcels that remain uninspected, for which you will enter these codes and hyperlink a photograph to complete the rapid damage assessment for this area.

1   Export the selected high risk Berkeley_Parcels to a new shapefile and save as **FL_Berkeley_ Parcels_HR**, with FL being your initials, in **\ESRIPress\GISTHS\MYGISTHS_Work**.

2   Drag this new data layer below the **FL_Land_Liq_PAccel_GE46** data layer, and turn off the Berkeley_Parcel, FL_Land_Liq_Paccel, and M73_Full_Hayward_Rodgers_Faults data layers.

3   Change the symbology of FL_Land_Liq_PAccel_GE46 to Cherrywood Brown, and increase its transparency to 50%.

4   Zoom to the block of parcels in Berkeley in the high risk zone.

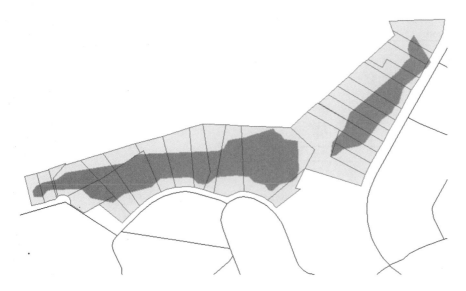

5   Label the Berkeley_StreetCenterLines data layer using the STR_NAM field.

6   Label the FL_Berkeley_Parcels_HR data layer using the I_STRNO field.

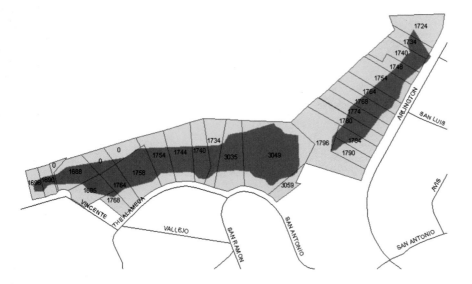

7 Change the symbology of the FL_Berkeley_Parcels_HR data layer to Categories showing the Unique Values of the Posting field. Select gray 40% for unlabeled parcels; green for G, red for R, and yellow for Y. Uncheck <all other values>.

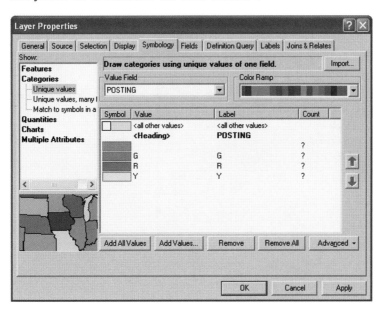

8 Click OK.

9 Clear selected features.

10 Turn off the BAUA_Counties data layer, and turn on the Berkeley_Parcels data layer to add the parcels outside of the hazard zone back to the map.

The map display shows the posting designation of the parcels that have been inspected and tagged. There are parcels on the block that have not yet been inspected. You will now select these parcels and enter assessment data to indicate the postevent status of the property.

Enter damage assessment data

1   Set FL_Berkeley_Parcels_HR as the only selectable layer.

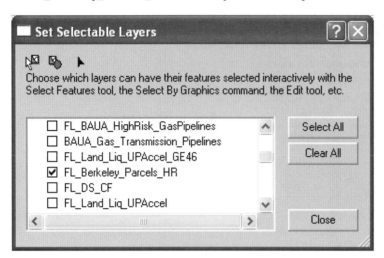

2   Use the Select Features tool [icon] to click on the parcel at 1768 Arlington.

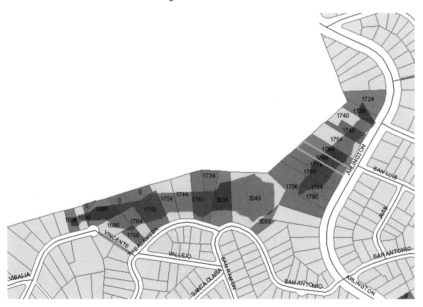

3   Open the FL_Berkeley_Parcels_HR attribute table and click Selected to see only this parcel.

4 Scroll to the right of this table to see the damage assessment fields.

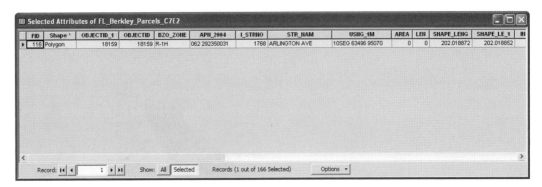

5 From the Editor tool bar, select Start Editing.

6 Select the **\ESRIPress\GISTHS\MYGISTHS_Work** folder where the FL_Berkeley_Parcels_HR shapefile is located.

7 Click OK.

8 Enter the following values into the table:

	1768
INSP_ID	**ABC998**
INSP_DATE	**8/19/07**
NUM_STORIE	**2**
FOOTPRINT_	**1800**
NUM_RESUNI	**1**
TYPE_CONST	**WF**
PRIM_OCCUP	**D**
COND_COLLA	**MN**
COND_LEAN	**MN**
COND_RACK	**MN**
COND_CHIMN	**MN**
COND_GROUN	**MN**
POSTING	**G**
F_ACTION	**N**
PCNT_DAMAG	**10**

9 Press Enter when complete and return to the map to see that this parcel is now colored green to show the posting designation.

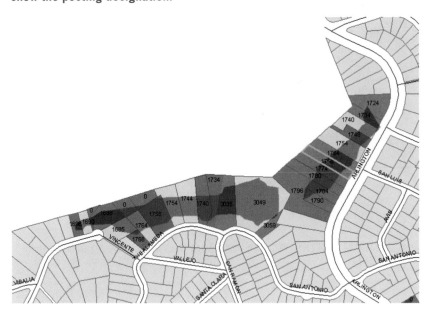

10 Enter the following values into the table for the remaining high-risk parcels on this block:

	1774	1780	1784	1790	1796
INSP_ID	**ABC998**	**ABC998**	**ABC998**	**ABC998**	**ABC998**
INSP_DATE	**8/19/07**	**8/19/07**	**8/19/07**	**8/19/07**	**8/19/07**
NUM_STORIE	**1**	**2**	**2**	**1**	**1**
FOOTPRINT_	**2000**	**1800**	**1700**	**1500**	**3000**
NUM_RESUNI	**1**	**1**	**1**	**1**	**1**
TYPE_CONST	**RF**	**WF**	**WF**	**RF**	**WF**
PRIM_OCCUP	**D**	**D**	**D**	**D**	**D**
COND_COLLA	**MD**	**SV**	**SV**	**MD**	**MN**
COND_LEAN	**MD**	**SV**	**SV**	**MD**	**MN**
COND_RACK	**MD**	**SV**	**SV**	**MD**	**MN**
COND_CHIMN	**MD**	**SV**	**SV**	**MN**	**MD**
COND_GROUN	**MD**	**SV**	**SV**	**SV**	**MN**
POSTING	**Y**	**R**	**R**	**Y**	**G**
F_ACTION	**DS**	**B,DS,DG**	**B,DS,DG**	**B,DG**	**N10**
PCNT_DAMAG	**30**	**70**	**80**	**60**	

11 Close the attribute table to view the posting designations of these parcels in this high-risk area.

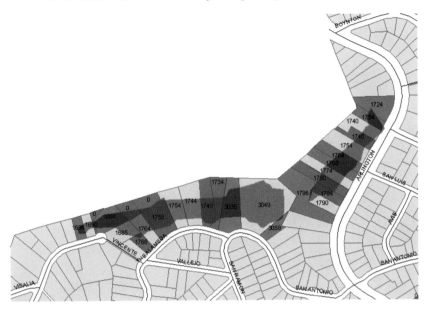

12 Stop and save edits, and clear selected features.

13 Save the map document when complete.

Review parcel recovery status maps

Now that each parcel in this area has been assessed for damage, a series of maps can be generated to assist in developing a recovery plan. Maps showing the parcels requiring further actions, percent damage, and the condition of the ground slope are useful to emergency operations personnel and recovery planners who allocate resources to secure and rebuild structures and seek disaster assistance from state and federal agencies.

1 Change the symbology of the parcels data layer to show what further actions are required for the inspected parcels.

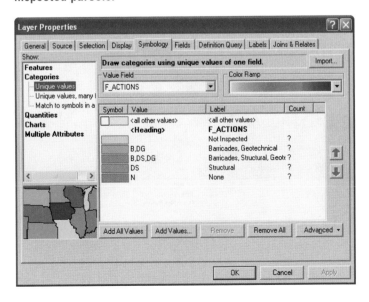

EXERCISES

2 Click OK.

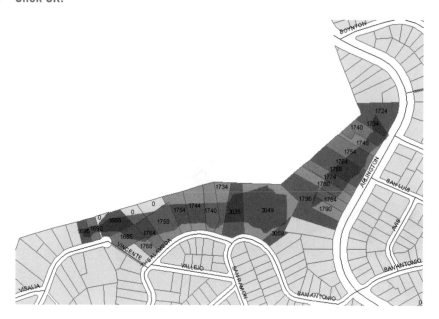

The map now displays the inspected parcels showing what further actions are required to return them to a safe condition for occupancy. Inspectors can use this map to allocate resources in the field to undertake detailed structural and geotechnical evaluations as required.

3 Create a map layout showing the further actions required for these inspected parcels. Select by attributes the parcels with INSP_DATE GT 08/19/07 to include a report in the layout listing the APN number, address, posting, and action for only these parcels.

Berkeley Rapid Damage Assessment

Further Actions Required

YOUR TURN

Create a new map layout showing the percent damage of the inspected parcels.

Hyperlink photos to maps

When inspectors are in the field, they often take ground view photographs to capture a visual record of the site condition. ArcMap enables the user to hyperlink these images to the map to display as a feature is clicked on. This is a very useful tool to emergency operations personnel and recovery planners, providing an additional piece of information to verify the status of a structure or surrounding ground condition.

1   Change the symbology of the parcels data layer to show the condition of collapse by mapping the COND_COLLA field. Change the sort order of the values as shown below.

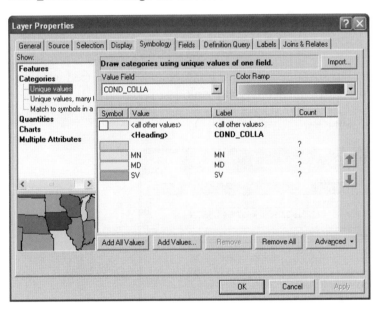

2   Click OK.

The parcels where the structure has sustained only minimal collapse are mapped in pale green; moderate collapse in pale yellow; and severe collapse in pink.

3   Use the Identify Tool to click on the parcel at 1748 Arlington.

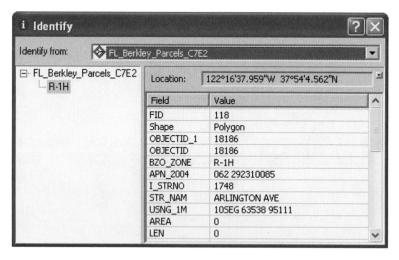

The Identify window appears listing all of the attributes of this parcel.

4   Right-click the R-1H field in the left window pane and select Add Hyperlink.

5   Browse to **\ESRIPress\GISTHS_C7** and select 1748.jpg.

6   Click OK.

7   Close the Identify window.

8   From the Tools toolbar, click the Hyperlink Tool.

The hyperlinked feature now has a blue outline around it.

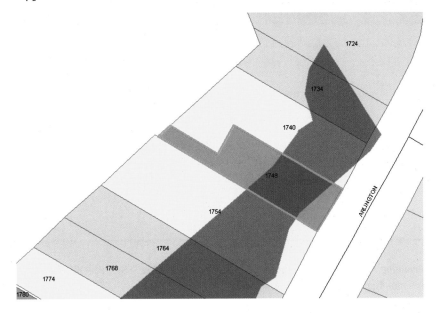

9 Use the Hyperlink Tool to click on the parcel at 1748 Arlington.

1748.jpg — Zoom:

J.K. Nakata, U.S. Geological Survey[17]

A photograph of the site is opened in an image viewer. This information is very useful to inspectors and recovery planners as they record the postevent conditions to determine the safety of the structure and surrounding ground conditions. The hyperlink function is not limited to photographs. Other documents, including word processing files, Web site URLs, PowerPoint presentation files, CAD drawings, and scanned images can also be hyperlinked to a feature for enhanced document support.

YOUR TURN

Hyperlink the remaining six photographs labeled by street address in the \ESRIPress\GISTHS_C7 folder to the corresponding parcel features on the map.

Summary

In this final chapter of the tutorial, you explored and applied the tools and analysis that GIS brings to the *Recover* Mission Area of homeland security planning and operations. Recover includes the immediate implementation of actions to restore essential services to a community after a major disaster or event. Using the National Planning Scenario 9, which focuses on a major earthquake in an urban area, you incorporated the USGS ShakeMaps and California Geological Survey Landslide and Liquefaction Hazard Zone Maps into the GIS to determine the locations in the Bay Area expected to experience the greatest impact from an earthquake of vast proportions. Within these areas, you used GIS to identify segments of high-pressure gas transmission lines that are at the greatest risk of rupture; traced electrical power disruptions suffered along the fault line; and prepared reports and maps of parcels experiencing outages. Knock-on effects of these outages were identified, illustrating the effects on other infrastructure, such as other public utilities, hospitals, churches and schools that are critical to a recovery effort. This information is essential to emergency operations personnel as they coordinate efforts with utility workers to restore service to impacted residents and services. Once lifeline restoration is under way, focus shifts to the assessment of damage incurred by individual homeowners. You integrated industry-standard damage assessment tools into the GIS to enable rapid and efficient evaluation of properties located in the greatest hazard zones. Hyperlinking photos to the GIS enhanced the functionality of the GIS by providing a visual record of assessed properties.

All of the GIS tools and analyses you practiced in this tutorial greatly advance our efforts to safeguard the security of our nation. Integrating these tools into homeland security planning and operations is essential to a coordinated and interoperable effort to build a well-designed and maintained geospatial infrastructure to support the public safety and well being of our nation.

Notes

1. DHS *National Planning Scenarios*, Version 20.1, April 2005, p. vii

2. DHS Homeland Security Programs and Activities, *http://www.usaha.org/committees/aem/presentations 2006/SAHA-DHS2006.pdf*, p. 12

3. DHS *Target Capabilities List*, Version 1.1, May 23, 2005, p. 152.

4. Ibid., p. 156.

5. Ibid., p. 156.

6. DHS *National Planning Scenarios*, Version 20.1, April 2005, p. 9-1

7. *http://earthquake.usgs.gov/learning/faq.php?categoryID=11&faqID=95* and *http://earthquake.usgs.gov/learning/glossary.php?termID=150*.

8. ABAG Earthquake Maps and Information, *http://quake.abag.ca.gov/*

9. An overview of the earthquake scenario program can be found at Bay Area Earthquake Scenarios, *http://seismo.berkeley.edu/hayward/scenarios.html*

10. DHS *National Planning Scenarios*, Version 20.1, April 2005, p. 9-6

11. *http://earthquake.usgs.gov/eqcenter/shakemap/list.php?s=1&n=nc*.

12. ArcFM, a powerful extension of ESRI's ArcGIS 9 platform, is a complete enterprise GIS solution for editing, modeling, maintenance, and management of facility information for electric, gas, and water/wastewater utilities.

13. Because of the sensitive nature of this data, the utility data layers used in this exercise are either in the public domain, or fictitious data layers created for demonstration purposes only. To ensure public safety, homeland security personnel would have access to the locations of these infrastructure features through a regulated cooperative data sharing agreement directly with the utility companies that own and operate these facilities.

14. Meehan, Bill. _Empowering Electric and Gas Utilities with GIS_. ESRI Press. Redlands, California. 2007. p. 105.

15. All of the street addresses and parcel numbers in this data layer have been randomly modified to protect the privacy and security of property owners in the city of Berkeley.

16. ATC-20-2, _Addendum to the ATC-20 Postearthquake Building Safety Evaluation Procedures_, prepared by the Applied Technology Council for the National Science Foundation, Redwood City, California, 1995. _www.atcouncil.org/fandp.shtml_

17. All photographs used in this exercise are from the USGS Earthquake Photo Collection _(http://earthquake.usgs.gov/learning/photos.php)_ and used for demonstration purposes only. They are not of the actual sites to which they are linked.

Appendix A
Data license agreement

Important: Read carefully before opening the sealed media package

Environmental Systems Research Institute Inc. (ESRI), is willing to license the enclosed data and related materials to you only upon the condition that you accept all of the terms and conditions contained in this license agreement. Please read the terms and conditions carefully before opening the sealed media package. By opening the sealed media package, you are indicating your acceptance of the ESRI License Agreement. If you do not agree to the terms and conditions as stated, then ESRI is unwilling to license the data and related materials to you. In such event, you should return the media package with the seal unbroken and all other components to ESRI.

ESRI license agreement

This is a license agreement, and not an agreement for sale, between you (Licensee) and Environmental Systems Research Institute Inc. (ESRI). This ESRI License Agreement (Agreement) gives Licensee certain limited rights to use the data and related materials (Data and Related Materials). All rights not specifically granted in this Agreement are reserved to ESRI and its Licensors.

Reservation of Ownership and Grant of License: ESRI and its Licensors retain exclusive rights, title, and ownership to the copy of the Data and Related Materials licensed under this Agreement and, hereby, grant to Licensee a personal, nonexclusive, nontransferable, royalty-free, worldwide license to use the Data and Related Materials based on the terms and conditions of this Agreement. Licensee agrees to use reasonable effort to protect the Data and Related Materials from unauthorized use, reproduction, distribution, or publication.

Proprietary Rights and Copyright: Licensee acknowledges that the Data and Related Materials are proprietary and confidential property of ESRI and its Licensors and are protected by United States copyright laws and applicable international copyright treaties and/or conventions.

Permitted Uses: Licensee may install the Data and Related Materials onto permanent storage device(s) for Licensee's own internal use.

Licensee may make only one (1) copy of the original Data and Related Materials for archival purposes during the term of this Agreement unless the right to make additional copies is granted to Licensee in writing by ESRI.

Licensee may internally use the Data and Related Materials provided by ESRI for the stated purpose of GIS training and education.

Uses Not Permitted: Licensee shall not sell, rent, lease, sublicense, lend, assign, time-share, or transfer, in whole or in part, or provide unlicensed Third Parties access to the Data and Related Materials or portions of the Data and Related Materials, any updates, or Licensee's rights under this Agreement.

Licensee shall not remove or obscure any copyright or trademark notices of ESRI or its Licensors.

Term and Termination: The license granted to Licensee by this Agreement shall commence upon the acceptance of this Agreement and shall continue until such time that Licensee elects in writing to discontinue use of the Data or Related Materials and terminates this Agreement. The Agreement shall automatically terminate without notice if Licensee fails to comply with any provision of this Agreement. Licensee shall then return to ESRI the Data and Related Materials. The parties hereby agree that all provisions that operate to protect the rights of ESRI and its Licensors shall remain in force should breach occur.

Disclaimer of Warranty: The Data and Related Materials contained herein are provided "as-is," without warranty of any kind, either express or implied, including, but not limited to, the implied warranties of merchantability, fitness for a particular purpose, or noninfringement. ESRI does not warrant that the Data and Related Materials will meet Licensee's needs or expectations, that the use of the Data and Related Materials will be uninterrupted, or that all nonconformities, defects, or errors can or will be corrected. ESRI is not inviting reliance on the Data or Related Materials for commercial planning or analysis purposes, and Licensee should always check actual data.

Data Disclaimer: The Data used herein has been derived from actual spatial or tabular information. In some cases, ESRI has manipulated and applied certain assumptions, analyses, and opinions to the Data solely for educational training purposes. Assumptions, analyses, opinions applied, and actual outcomes may vary. Again, ESRI is not inviting reliance on this Data, and the Licensee should always verify actual Data and exercise their own professional judgment when interpreting any outcomes.

Limitation of Liability: ESRI shall not be liable for direct, indirect, special, incidental, or consequential damages related to Licensee's use of the Data and Related Materials, even if ESRI is advised of the possibility of such damage.

No Implied Waivers: No failure or delay by ESRI or its Licensors in enforcing any right or remedy under this Agreement shall be construed as a waiver of any future or other exercise of such right or remedy by ESRI or its Licensors.

Order for Precedence: Any conflict between the terms of this Agreement and any FAR, DFAR, purchase order, or other terms shall be resolved in favor of the terms expressed in this Agreement, subject to the government's minimum rights unless agreed otherwise.

Export Regulation: Licensee acknowledges that this Agreement and the performance thereof are subject to compliance with any and all applicable United States laws, regulations, or orders relating to the export of data thereto. Licensee agrees to comply with all laws, regulations, and orders of the United States in regard to any export of such technical data.

Severability: If any provision(s) of this Agreement shall be held to be invalid, illegal, or unenforceable by a court or other tribunal of competent jurisdiction, the validity, legality, and enforceability of the remaining provisions shall not in any way be affected or impaired thereby.

Governing Law: This Agreement, entered into in the County of San Bernardino, shall be construed and enforced in accordance with and be governed by the laws of the United States of America and the State of California without reference to conflict of laws principles. The parties hereby consent to the personal jurisdiction of the courts of this county and waive their rights to change venue.

Entire Agreement: The parties agree that this Agreement constitutes the sole and entire agreement of the parties as to the matter set forth herein and supersedes any previous agreements, understandings, and arrangements between the parties relating hereto.

Appendix B
Installing the data and software

GIS Tutorial for Homeland Security includes two DVDs at the back of the book. One DVD contains exercise data and instructional resource material. The other DVD contains ArcGIS 9.3 Desktop (ArcView license, Single Use, 180-day trial) software. Installation of the ArcGIS Desktop software DVD with extensions takes approximately 30 minutes and requires at least 1.5 GB of hard-disk space. Installation times will vary with your computer's speed and available memory.

If you already have a licensed copy of ArcGIS Desktop 9.3 installed on your computer (or accessible through a network), do not install the software DVD. Use your licensed software to do the exercises in this book. If you have an older version of ArcGIS installed on your computer, you must uninstall it before you can install the software DVD that comes with this book.

The exercises in this book only work with ArcGIS 9.3 or higher.

Installing the exercise data

Follow the steps below to install the exercise data.

1   Put the data DVD in your computer's DVD drive. A splash screen will appear.

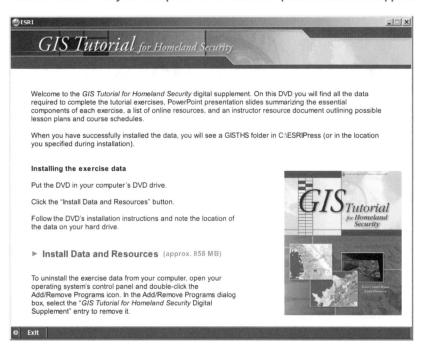

2   Read the welcome, then click the Install Exercise Data link. This launches the InstallShield Wizard.

3   Click Next. Read and accept the license agreement terms, then click Next.

4   Accept the default installation folder. We recommend that you do not choose an alternate location.

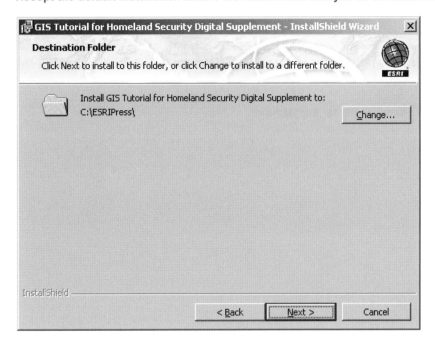

5   Click Next. The installation will take a few moments. When the installation is complete, you will
     see the following message.

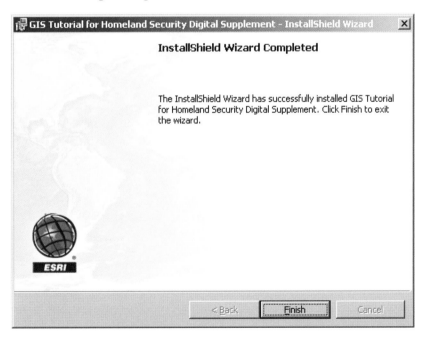

6   Click Finish. The exercise data is installed on your computer in a folder called GISTHS.

Uninstalling the data and resources

To uninstall the data and resources from your computer, open your operating system's control panel and double-click the Add/Remove Programs icon. In the Add/Remove Programs dialog box, select the following entry and follow the prompts to remove it:

GIS Tutorial for Homeland Security-Digital Supplement

Installing the software

The ArcGIS software included on this DVD is intended for educational purposes only. Once installed and registered, the software will run for 180 days. The software cannot be reinstalled nor can the time limit be extended. It is recommended that you uninstall this software when it expires.

Follow the steps below to install the software.

1   Put the software DVD in your computer's DVD drive. A splash screen will appear.

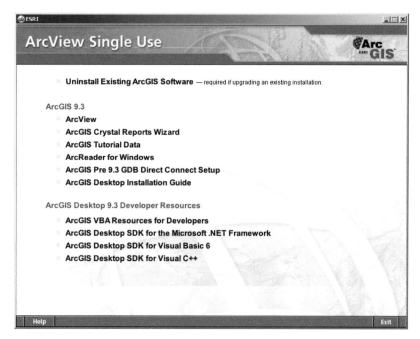

2   Click the ArcGIS ArcView installation option. On the Startup window, click Install ArcGIS Desktop. This will launch the Setup wizard.

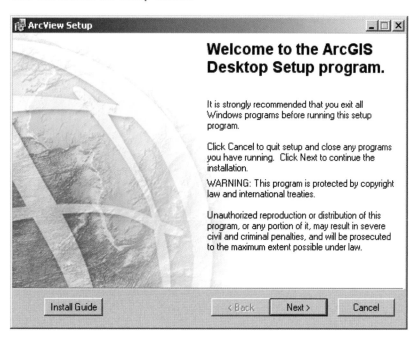

3 Read the Welcome, then click Next.

4 Read the license agreement. Click "I accept the license agreement" and click Next.

5 The default installation type is Typical. You must choose the Complete install, which will add extension products that are used in the book. Click the button next to Complete install.

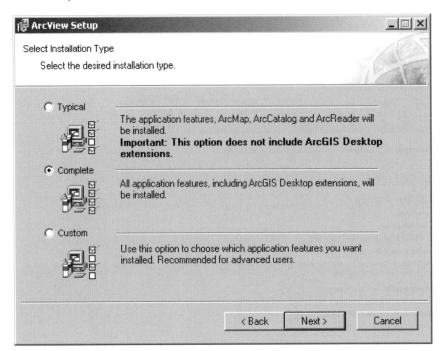

6 Click Next. Accept the default installation folder or click Browse and navigate to the drive or folder location where you want to install the software.

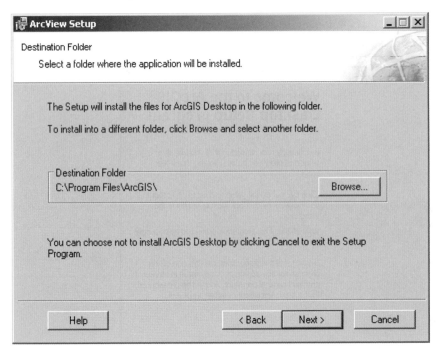

7   Click Next. Accept the default installation folder or navigate to the drive or folder where you want to install Python, a scripting language used by some ArcGIS geoprocessing functions. (You won't see this panel if you already have Python installed.) Click Next.

8   The installation paths for ArcGIS and Python are confirmed. Click Next. The software will take some time to install on your computer. When the installation is finished, you see the following message:

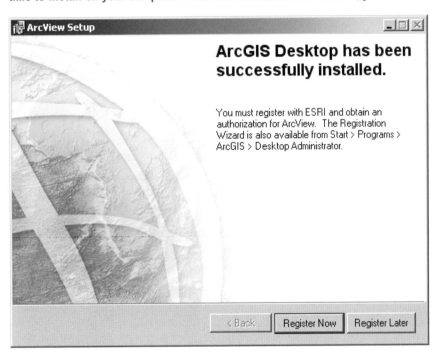

9   Click Register Now and follow the registration process. The registration code is located at the bottom of the software DVD jacket in the back of the book. Be sure to register the 3D Analyst and Spatial Analyst extensions as well as any other extensions you might be interested in exploring.

If you have questions or encounter problems during the installation process, or while using this book, please use the resources listed below. (The ESRI Technical Support Department does not answer questions regarding the ArcGIS 9 software DVD, the *GIS Tutorial for Homeland Security Digital Supplement*, or the contents of the book itself.)

+ To resolve problems with the trial software or exercise data, or to report mistakes in the book, send an e-mail to ESRI workbook support at *workbook-support@esri.com*.

+ To stay informed about exercise updates, FAQs, and errata, visit the book's Web page at *www.esri.com/gistutorialhomeland*.

Uninstalling the software

To uninstall the software from your computer, open your operating system's control panel and double-click the Add/Remove Programs icon. In the Add/Remove Programs dialog box, select the following entry and follow the prompts to remove it:

ArcGIS Desktop

Appendix C
Data sources

Chapter 1 data sources include:

\GISTHS\GISTHS_C1\USA_Cities.shp, from ESRI Data and Maps, 2006, courtesy of National Atlas of the United States.
\GISTHS\GISTHS_C1\USA_Interstates.shp, from ESRI Data and Maps, 2006, courtesy of U.S. Bureau of Transportation Statistics.
\GISTHS\GISTHS_C1\USA_States.shp, from ESRI Data and Maps, 2006, courtesy of ArcUSA, U.S. Census, and ESRI(Pop2005 field).

Chapter 2 data sources include:

\GISTHS\GISTHS_C2\USA_Cities.shp, from ESRI Data and Maps, 2006, courtesy of National Atlas of the United States.
\GISTHS\GISTHS_C2\USA_Counties.shp, from ESRI Data and Maps, 2006, courtesy of ArcUSA, U.S. Census, and ESRI (Pop2005 field).
\GISTHS\GISTHS_C2\USA_Interstates.shp, from ESRI Data and Maps, 2006, courtesy of U.S. Bureau of Transportation Statistics.
\GISTHS\GISTHS_C2\USA_States.shp, from ESRI Data and Maps, 2006, courtesy of ArcUSA, U.S. Census, and ESRI (Pop2005 field).

Chapter 3 data sources include:

\GISTHS\GISTHS_C3\cfcc.dbf, from ESRI Data and Maps, 2006, courtesy of U.S. Census.
\GISTHS\GISTHS_C3\BAUA_Airports.shp, courtesy of U.S. Census Bureau TIGER.
\GISTHS\GISTHS_C3\BAUA_Landpts.shp, courtesy of U.S. Census Bureau TIGER.
\GISTHS\GISTHS_C3\BAUA_Railroads.shp, courtesy of U.S. Census Bureau TIGER.
\GISTHS\GISTHS_C3\USA_Counties.shp, from ESRI Data and Maps, 2006, courtesy of ArcUSA, U.S. Census, and ESRI(Pop2005 field).
\GISTHS\GISTHS_C3\USA_Places.shp, from ESRI Data and Maps, 2006, courtesy of Tele Atlas and U.S. Census.
\GISTHS\GISTHS_C3\USA_States.shp, from ESRI Data and Maps, 2006, courtesy of ArcUSA, U.S. Census, and ESRI(Pop2005 field).
\GISTHS\GISTHS_C3\Downloaded_Data_C3\BAUA_BTSRoads\30835236.shp, data available from U.S. Geological Survey, EROS Data Center, Sioux Falls, South Dakota.

\GISTHS\GISTHS_C3\Downloaded_Data_C3\BAUA_DEM\baua_dem, data available from U.S. Geological Survey, EROS Data Center, Sioux Falls, South Dakota.

\GISTHS\GISTHS_C3\Downloaded_Data_C3\BAUA_GeogNames\CA_DECI.txt, courtesy of U.S. Geological Survey.

\GISTHS\GISTHS_C3\Downloaded_Data_C3\BAUA_LandCover\05567535, data available from U.S. Geological Survey, EROS Data Center, Sioux Falls, South Dakota.

\GISTHS\GISTHS_C3\Downloaded_Data_C3\BAUA_Landmarks\tgr06001lpt.shp, courtesy of U.S. Census Bureau TIGER.

\GISTHS\GISTHS_C3\Downloaded_Data_C3\BAUA_Landmarks\tgr06013lpt.shp, courtesy of U.S. Census Bureau TIGER.

\GISTHS\GISTHS_C3\Downloaded_Data_C3\BAUA_Landmarks\tgr06041lpt.shp, courtesy of U.S. Census Bureau TIGER.

\GISTHS\GISTHS_C3\Downloaded_Data_C3\BAUA_Landmarks\tgr06075lpt.shp, courtesy of U.S. Census Bureau TIGER.

\GISTHS\GISTHS_C3\Downloaded_Data_C3\BAUA_Landmarks\tgr06081lpt.shp, courtesy of U.S. Census Bureau TIGER.

\GISTHS\GISTHS_C3\Downloaded_Data_C3\BAUA_Landmarks\tgr06087lpt.shp, courtesy of U.S. Census Bureau TIGER.

\GISTHS\GISTHS_C3\Downloaded_Data_C3\BAUA_Landmarks\tgr06095lpt.shp, courtesy of U.S. Census Bureau TIGER.

\GISTHS\GISTHS_C3\Downloaded_Data_C3\BAUA_Landmarks\tgr06097lpt.shp, courtesy of U.S. Census Bureau TIGER.

\GISTHS\GISTHS_C3\Downloaded_Data_C3\BAUA_Landmarks\tgr06099lpt.shp, courtesy of U.S. Census Bureau TIGER.

\GISTHS\GISTHS_C3\Downloaded_Data_C3\BAUA_Landmarks\tgr06001lpy.shp, courtesy of U.S. Census Bureau TIGER.

\GISTHS\GISTHS_C3\Downloaded_Data_C3\BAUA_Landmarks\tgr06013lpy.shp, courtesy of U.S. Census Bureau TIGER.

\GISTHS\GISTHS_C3\Downloaded_Data_C3\BAUA_Landmarks\tgr06041lpy.shp, courtesy of U.S. Census Bureau TIGER.

\GISTHS\GISTHS_C3\Downloaded_Data_C3\BAUA_Landmarks\tgr06055lpy.shp, courtesy of U.S. Census Bureau TIGER.

\GISTHS\GISTHS_C3\Downloaded_Data_C3\BAUA_Landmarks\tgr06075lpy.shp, courtesy of U.S. Census Bureau TIGER.

\GISTHS\GISTHS_C3\Downloaded_Data_C3\BAUA_Landmarks\tgr06081lpy.shp, courtesy of U.S. Census Bureau TIGER.

\GISTHS\GISTHS_C3\Downloaded_Data_C3\BAUA_Landmarks\tgr06087lpy.shp, courtesy of U.S. Census Bureau TIGER.

\GISTHS\GISTHS_C3\Downloaded_Data_C3\BAUA_Landmarks\tgr06095lpy.shp, courtesy of U.S. Census Bureau TIGER.

\GISTHS\GISTHS_C3\Downloaded_Data_C3\BAUA_Landmarks\tgr06097lpy.shp, courtesy of U.S. Census Bureau TIGER.

\GISTHS\GISTHS_C3\Downloaded_Data_C3\BAUA_Landmarks\tgr06099lpy.shp, courtesy of U.S. Census Bureau TIGER.

\GISTHS\GISTHS_C3\Downloaded_Data_C3\BAUA_NHD1805\NHDH1805.mdb, courtesy of U.S. Geological Survey.

\GISTHS\GISTHS_C3\Downloaded_Data_C3\BAUA_Orthoimagery\21319435, data available from U.S. Geological Survey, EROS Data Center, Sioux Falls, South Dakota.

Chapter 4 data sources include:

\GISTHS\GISTHS_C4\US_States_FIPS.xls, courtesy of U.S. Department of Commerce/National Institute of Standards and Technology.
\GISTHS\GISTHS_C4\NA_Airports_DAFIF.shp, courtesy of Pacific Disaster Center (pdc.org).
\GISTHS\GISTHS_C4\USA_Interstates.shp, from ESRI Data and Maps, 2006, courtesy of U.S. Bureau of Transportation Statistics.
\GISTHS\GISTHS_C4\USA_States.shp, from ESRI Data and Maps, 2006, courtesy of ArcUSA, U.S. Census, ESRI (Pop2005 field).

Chapter 5 data sources include:

\GISTHS\GISTHS_C5\Petrol_Tanks.shp, courtesy of Berkeley Geo Research Group.
\GISTHS\GISTHS_C5\PLUME_COBALT_HYDROCARBONYL.shp, courtesy of Berkeley Geo Research Group.
\GISTHS\GISTHS_C5\Downloaded_Data_C5\tgr06000sf1blk.dbf, courtesy of U.S. Census Bureau TIGER.
\GISTHS\GISTHS_C5\Downloaded_Data_C5\tgr06013blk00.shp, courtesy of U.S. Census Bureau TIGER.

Chapter 6 data sources include:

\GISTHS\GISTHS_C6\JacksonCo_MEDS.gdb\jack_elev, data available from U.S. Geological Survey, EROS Data Center, Sioux Falls, South Dakota.
\GISTHS\GISTHS_C6\JacksonCo_MEDS.gdb\jack_lulc, data available from U.S. Geological Survey, EROS Data Center, Sioux Falls, South Dakota.
\GISTHS\GISTHS_C6\JacksonCo_MEDS.gdb\JacksonCo_County, courtesy of U.S. Census Bureau TIGER.
\GISTHS\GISTHS_C6\JacksonCo_MEDS.gdb\JacksonCo_GeogNames, courtesy of U.S. Geological Survey.
\GISTHS\GISTHS_C6\JacksonCo_MEDS.gdb\JacksonCo_Hydrog_Streams, courtesy of U.S. Census Bureau TIGER.
\GISTHS\GISTHS_C6\JacksonCo_MEDS.gdb\JacksonCo_Hydrog_WatBodies, courtesy of U.S. Census Bureau TIGER.
\GISTHS\GISTHS_C6\JacksonCo_MEDS.gdb\JacksonCo_Powerlines, courtesy of U.S. Census Bureau TIGER.
\GISTHS\GISTHS_C6\JacksonCo_MEDS.gdb\JacksonCo_Streets, courtesy of U.S. Census Bureau TIGER.
\GISTHS\GISTHS_C6\jack_fld1, courtesy of Berkeley Geo Research Group.
\GISTHS\GISTHS_C6\jack_fld2, courtesy of Berkeley Geo Research Group.
\GISTHS\GISTHS_C6\jack_fld3, couretsy of Berkeley Geo Research Group.
\GISTHS\GISTHS_C6\JacksonCo_MP.csv, courtesy of Berkeley Geo Research Group.

Chapter 7 data sources include:

\GISTHS\GISTHS_C7\Berkeley_MEDS.gdb\Berkeley_Parcels, courtesy of City of Berkeley.
\GISTHS\GISTHS_C7\Berkeley_MEDS.gdb\Berkeley_StreetCenterLines, courtesy of the GIS Center, University of California, Berkeley.
\GISTHS\GISTHS_C7\Berkeley_MEDS.gdb\power, courtesy of Berkeley Geo Research Group.
\GISTHS\GISTHS_C7\1734.jpg, courtesy of U.S. Geological Survey.
\GISTHS\GISTHS_C7\1740.jpg, courtesy of U.S. Geological Survey.
\GISTHS\GISTHS_C7\1748.jpg, courtesy of U.S. Geological Survey.
\GISTHS\GISTHS_C7\1768.jpg, courtesy of U.S. Geological Survey.
\GISTHS\GISTHS_C7\1780.jpg, courtesy of U.S. Geological Survey.
\GISTHS\GISTHS_C7\1784.jpg, courtesy of U.S. Geological Survey.
\GISTHS\GISTHS_C7\USGS_BA_Faults.shp, courtesy of U.S. Geological Survey.
\GISTHS\GISTHS_C7\USGS_M73_Full_Hayward_Rodgers_Faults.shp, courtesy of U.S. Geological Survey.
ATC-20 Building Safety Evaluation Forms, courtesy of Advanced Technology Council, http://www.atcouncil.org/fandp.shtml
\GISTHS\GISTHS_C7\BAUA_Gas_Transmission_Pipelines.shp, courtesy of Berkeley Geo Research Group.

Appendix D
Online resources

Applied Technology Council
http://www.atcouncil.org/fandp.shtml

The Applied Technology Council (ATC) is a nonprofit, tax-exempt corporation established in 1973 through the efforts of the Structural Engineers Association of California. ATC's mission is to develop and promote state-of-the-art, user-friendly engineering resources and applications for mitigating the effects of natural and other hazards on the built environment. ATC also identifies and encourages research and develops consensus on structural engineering issues in a nonproprietary format.

DHS List of Fiscal Year 2006 Urban Areas Security Initiative (UASI) eligible applicants
http://www.dhs.gov/xlibrary/assets/FY06_UASI_Eligibility_List.pdf

The FY 2006 UASI program provides financial assistance to address the unique multidiscipline planning, operations, equipment, training, and exercise needs of high-threat, high-density Urban Areas, and to assist them in building and sustaining capabilities to prevent, protect against, respond to, and recover from threats or acts of terrorism. This site includes a table listing how each candidate urban area in the United States is to be geographically captured to delineate the extent of area covered by UASI funding.

DHS Fiscal Year 2007 Homeland Security Grant Program
http://www.dhs.gov/xlibrary/assets/grants_st-local_fy07.pdf

Full PDF document of U.S. Department of Homeland Security grant programs.

Digital aeronautical flight information file
http://www.pdc.org/geodata/world/

A list of worldwide aeronautical and flight information databases available for download.

Computer-Aided Management of Emergency Operations (CAMEO)

http://www.epa.gov/oem/content/cameo/index.htm

CAMEO is a system of software applications used widely to plan for and respond to chemical emergencies. It is one of the tools developed by U.S. Environmental Protection Agency's Office of Emergency Management and the National Oceanic and Atmospheric Administration Office of Response and Restoration to assist front-line chemical emergency planners and responders. They can use CAMEO to access, store, and evaluate information critical for developing emergency plans. In addition, CAMEO supports regulatory compliance by helping users meet the chemical inventory reporting requirements of the Emergency Planning and Community Right-to-Know Act (EPCRA, also known as SARA Title III).

ESRI geographic data portal

www.esri.com/data/download/census2000_tigerline/index.html

GIS users can download Census 2000 TIGER/Line Data in shapefile format for an area of interest. Users can choose multiple data layers for a single county or a single data layer for multiple counties and analyze them using GIS software such as ArcGIS and ArcGIS Explorer.

Geographic Names Information System

geonames.usgs.gov/domestic/download_data.htm

The Geographic Names Information System (GNIS) is the federal standard for geographic nomenclature. The U.S. Geological Survey developed the GNIS for the U.S. Board on Geographic Names as the official repository of domestic geographic names data; the official vehicle for geographic names use by all departments of the federal government; and the source for applying geographic names to federal electronic and printed products.

Mississippi Automated Resource Information System (MARIS)

http://www.maris.state.ms.us/HTM/DownloadData/County_Standard.html

MARIS serves as the legislative mechanism within Mississippi state government to provide for the systematic arrangement, availability, and use of digital natural and cultural resource information. MARIS encourages compatibility of GIS and data distribution within state government.

National Hydrography Dataset

http://nhd.usgs.gov/

The National Hydrography Dataset (NHD) is a comprehensive set of digital spatial data that contains information about surface water features such as lakes, ponds, streams, rivers, springs and wells. Within the NHD, surface water features are combined to form "reaches," which provide the framework for linking water-related data to the NHD surface water drainage network. These linkages enable the analysis and display of these water-related data in upstream and downstream order.

The National Map Seamless Server

http://seamless.usgs.gov/

The National Map Seamless Server is the ultimate location to explore and retrieve data. The U.S. Geological Survey (USGS) and the EROS Data Center (EDC) are committed to providing access to geospatial data through the National Map. An approach is to provide free downloads of national base layers, as well as other geospatial data layers.

Earthquake Hazards Program/Quaternary Fault and Fold Database Home
http://gldims.cr.usgs.gov/qfault/viewer.htm

This site contains information on faults and associated folds in the United States that are believed to be sources of greater-than-magnitude 6 earthquakes during the Quaternary (the past 1,600,000 years). Maps of these geologic structures are linked to detailed descriptions and references.

ShakeMaps
http://earthquake.usgs.gov/eqcenter/shakemap/

ShakeMap is a product of the U.S. Geological Survey Earthquake Hazards Program in conjunction with regional seismic network operators. ShakeMap sites provide near-real-time maps of ground motion and shaking intensity following significant earthquakes. These maps are used by federal, state, and local organizations, both public and private; for post-earthquake response and recovery; and public and scientific information, as well as for preparedness exercises and disaster planning.

USGS The Universal Transverse Mercator (UTM) Grid
http://erg.usgs.gov/isb/pubs/factsheets/fs07701.pdf

This site contains the fact sheet for the universal transverse Mercator grid that covers the conterminous 48 United States and comprises 10 zones.